T0212510

Lecture Notes in Computer Science　　10186

Commenced Publication in 1973
Founding and Former Series Editors:
Gerhard Goos, Juris Hartmanis, and Jan van Leeuwen

More information about this series at http://www.springer.com/series/7409

Fusheng Wang · Lixia Yao
Gang Luo (Eds.)

Data Management and Analytics for Medicine and Healthcare

Second International Workshop, DMAH 2016
Held at VLDB 2016
New Delhi, India, September 9, 2016
Revised Selected Papers

 Springer

Editors
Fusheng Wang
Stony Brook University
Stony Brook, NY
USA

Gang Luo
University of Utah
Salt Lake City, UT
USA

Lixia Yao
University of North Carolina at Charlotte
Charlotte, NC
USA

ISSN 0302-9743 ISSN 1611-3349 (electronic)
Lecture Notes in Computer Science
ISBN 978-3-319-57740-1 ISBN 978-3-319-57741-8 (eBook)
DOI 10.1007/978-3-319-57741-8

Library of Congress Control Number: 2017938565

LNCS Sublibrary: SL3 – Information Systems and Applications, incl. Internet/Web, and HCI

Printed on acid-free paper

This Springer imprint is published by Springer Nature
The registered company is Springer International Publishing AG
The registered company address is: Gewerbestrasse 11, 6330 Cham, Switzerland

Preface

In this volume we present the accepted contributions for the Second International Workshop on Data Management and Analytics for Medicine and Healthcare (DMAH 2016), held in New Delhi, India, in conjunction with the 42nd International Conference on Very Large Data Bases on September 5–9, 2016.

The goal of the DMAH workshop is to bring together people in the field cross-cutting information management and medical informatics to discuss innovative data management and analytics technologies highlighting end-to-end applications, systems, and methods to address problems in health care, public health, and everyday wellness, with clinical, physiological, imaging, behavioral, environmental, and -omic data as well as data from social media and the Web. It provides a unique opportunity for interaction between information management researchers and biomedical researchers in this interdisciplinary field.

For DMAH 2016, we received 11 papers. A rigorous, single-blind, peer-review selection procedure was adopted, resulting in seven accepted papers presented at the workshop. Each paper was reviewed by three members of the Program Committee, who were carefully selected for their knowledge and competence. As far as possible, papers were matched with the reviewer's particular interests and special expertise. The result of this careful process can be seen here in the high quality of the contributions published in this volume.

We would like to express our sincere thanks especially to the internationally renowned speakers who gave keynote talks at the workshops plenary sessions: Prof. Yuval Shahar of Ben-Gurion University, Beer Sheva, Israel; Prof. Adi V. Gundlapalli of VA Salt Lake City Health Care System and University of Utah School of Medicine, Salt Lake City, Utah, USA; and Prof. Supten Sarbadhikari of National Health Portal, India. We would like to thank the members of the Program Committee for their attentiveness, perseverance, and willingness to provide high-quality reviews.

March 2017

Fusheng Wang
Lixia Yao
Gang Luo

Organization

DMAH 2016

Workshop Chairs

Fusheng Wang	Stony Brook University, USA
Lixia Yao	University of North Carolina at Charlotte, USA
Gang Luo	University of Utah, USA

Program Committee

Syed Sibte Raza Abidi	Dalhousie University, Canada
Carlo Combi	University of Verona, Italy
Vassilis Cutsuridis	Foundation for Research and Technology – Hellas, Greece
Amarendra Das	Dartmouth College, USA
Anna Divoli	Pingar Research, New Zealand
Kerstin Denecke	Bern University of Applied Sciences, Switzerland
Dejing Dou	University of Oregon, USA
Peter Elkin	University of Buffalo, USA
David Greenhalgh	University of Strathclyde, UK
Jesús B. Alonso Hernández	University of Las Palmas de Gran Canaria, Spain
Guoqian Jiang	Mayo Clinic, USA
Ying Li	IBM T.J. Watson Research Center, USA
Jun Kong	Emory University, USA
Tahsin Kurc	Stony Brook University, USA
Ulf Leser	Humboldt-Universität zu Berlin, Germany
Yanhui Liang	Stony Brook University, USA
Fernando Martin-Sanchez	Weill Cornell Medicine, USA
Casey Lynnette Overby	Johns Hopkins School of Medicine, USA
Yuval Shahar	Ben-Gurion University, Israel
Xiang Li	Fudan University, China
Hua Xu	University of Texas Health Science Center, USA
Chao Yang	Amazon, USA
Lin Yang	University of Florida, USA
Zhe He	Florida State University, USA

Abstracts of Keynotes

High Yield Document Sets: An Information Extraction Strategy for Large Clinical Text Corpora

Adi V. Gundlapalli

VA Salt Lake City Health Care System, University of Utah School of Medicine,
Salt Lake City, UT, USA
adi.gundlapalli@hsc.utah.edu

Abstract. Extracting relevant concepts from clinical text corpora is an important strategy for quality improvement, operations, and research. With increasing use of electronic medical records and documentation by medical providers, the clinical text corpora needed to be processed is large. While it may be computationally feasible to process all available text documents, it would be beneficial to develop a strategy to identify high yield document sets for relevant concepts. Using principles of information foraging theory, we describe a method to identify high yield documents to enable processing of limited portions of large clinical text corpora for domain-specific information extraction. We present three use cases from real world examples using US Department of Veterans Affairs (VA) electronic medical record data. The VA is one of the largest health care systems in the US that treats nearly 6 million unique individuals every year; the data generated includes hundreds of millions of text notes from 83 million outpatient and 600,000 inpatient visits annually. The methods described are generalizable to electronic medical records from other health care systems.

eHealth Initiatives in Digital India

Suptendra Nath Sarbadhikari

Centre for Health Informatics, National Health Portal NIHFW, New Delhi
110067, India
supten@gmail.com

Abstract. The **National eHealth Authority (NeHA)** is being set up for facilitating the smooth adoption of eHealth in India. Following the putting up of the Concept Note in the public domain, and, subsequently after revising the Concept Note, based on the comments received from all the stakeholders, a National Consultation on National eHealth Authority (NeHA) was held on 4th April, 2016. Based on the deliberations of the Workshop, now a draft Bill/Act is being prepared for further processing.

An Integrated Health Information Platform (IHIP) is also being proposed so that Standards-compliant EHR systems can talk to each other and health information can be exchanged meaningfully.

All the healthcare facilities in India are now being provided a unique and permanent National Identification Number (NIN). This could be a starting point for linking data from various healthcare facilities throughout India.

As per the **Digital India** initiatives, the fifth pillar of eKranti includes Technology for Healthcare or eHealth having online medical consultation, online medical records, online medicine supply, and pan-India exchange for patient information.

For successful and meaningful information of health information, appropriate **Standards** are necessary to maintain interoperability between diverse Electronic Health Record (EHR) systems. The Ministry of Health and Family Welfare had recommended the Guidelines for EHR Standards in August 2013. Subsequently, India has become a country member of IHTSDO, which develops and maintains the SNOMED-CT. Now the Guidelines for EHR Standards are being revised and comments have been received on the proposed new edition. This will be notified soon.

Keywords: eHealth · Digital India · National eHealth Authority · Standards for Health Information Exchange

The Representation, Application, and Automated Discovery of Clinical Temporal-Abstraction Knowledge

Yuval Shahar

The Josef Erteschik Chair in Information Systems Engineering, Head, Medical Informatics Research Center, Department of Information Systems Engineering, Ben Gurion University, Beer Sheva 84105, Israel
yshahar@bgu.ac.il
http://www.ise.bgu.ac.il/faculty/shahar/

Monitoring, interpretation, and analysis of large amounts of time-stamped clinical data are subtasks that are at the core of multiple tasks important for medical care and research. Examples include the management of chronic patients using clinical guidelines, the retrospective assessment of the quality of the application of such a guideline, and the learning of new knowledge from analyzing the data regarding repeating patterns of measured data, and of meaningful abstractions derivable from these data. Such new knowledge can support additional tasks such as clustering, classification, and prediction.

My talk describes several conceptual and computational architectures developed over the past 20 years, mostly by my research teams at Stanford and Ben Gurion universities, for knowledge-based performance of these tasks, and highlights the complex and interesting relationships amongst them. Examples of such architectures include the IDAN and Momentum goal-directed and data-driven temporal-abstraction architectures, the KNAVE-II and VISITORS interactive-exploration frameworks for single and multiple longitudinal records, the KarmaLego temporal data mining methodology, and the ViTA-Lab framework, which integrates goal-directed and data-driven analysis of large numbers of multivariate time-oriented clinical data.

My talk also points out the progression, from individual-patient monitoring, diagnosis, and therapy, to multiple-patient aggregate analysis and research, and finally to the learning of new knowledge. This progression can be viewed as a positive-feedback loop, in which newly discovered knowledge can be exploited, on one hand, to improve the process of individual-patient management, and on the other hand, to learn additional meaningful (temporal) knowledge.

Abstracts of Invited Papers

Patient Records Retrieval System for Integrated Care in Treatment of Cervical Spine Defect

Yihan Deng[1,2] and Kerstin Denecke[2]

[1] Innovation Center Computer Assisted Surgery,
University of Leipzig, Leipzig, Germany
dengyihan@gmail.com
[2] Department of Medical Informatics,
Bern University of Applied Sciences, Biel, Switzerland
{yihan.deng,kerstin.denecke}@bfh.ch

Abstract. In clinical decision making, information on the treatment of patients that show similar medical conditions and symptoms to the current case, is one of most relevant information sources to create a good, evidence-based treatment plan. However, the retrieval of similar cases is still challenging and automatic support is missing. The reasons are two-fold: First, the query formulation is difficult since multiple criteria need to be selected and specified in short query phrases. Second, the discrete storage of multimedia patient records makes the retrieval and summary of a patient history extremely difficult. In this paper, we present a retrieval system for electronic health records (EHR). More specifically, a retrieval platform for EHRs for supporting clinical decision making in treatment of cervical spine defects with the information extracted from textual data of patient records is implemented as prototype. The patient cases are classified according to cervical spine defect classes, while the classification relies upon rules obtained from the corresponding defect classification schema and guidelines. In a retrospective study, the classifier is applied to clinical documents and the classification results are evaluated.

Building an i2b2-Based Integrated Data Repository for Cancer Research: A Case Study of Ovarian Cancer Registry

Na Hong[1,2], Zheng Li[1,3], Richard C. Kiefer[1], Melissa S. Robertson[1], Ellen L. Goode[1], Chen Wang[1], and Guoqian Jiang[1]

[1] Department of Health Sciences Research, Mayo Clinic, Rochester, MN, USA
{hong.na,li.zheng,kiefer.richard,robertson.melissa1,
egoode,wang.chen,jiang.guoqian}@mayo.edu
[2] Institute of Medical Information, Chinese Academy of Medical Sciences, Beijing, China
[3] Department of Gynecologic Oncology, The Third Affiliated Hospital of Kunming Medical University, Kunming, Yunnan, China

Abstract. In this study, we describe our preliminary efforts in building an i2b2-based integrated data repository that supports centralized data management for ovarian cancer clinical research, and discuss important lessons learnt that would inspire the evaluation and enhancement for future generic cancer-specific data repository. We collected multiple types of heterogeneous clinical data, including demographic, outcome, chemo-treatment and lab-test information for ovarian cancer. To better integrate different data types, we conducted data normalization procedures through reusing standard codes and creating mappings between local codes and standard vocabularies. We also developed the extract, transform and load (ETL) scripts to load the data into an i2b2 instance. Through further analytic practices, we evaluated major expectations of the systems according to common clinical research needs, including cohort query and identification, clinical data-based hypothesis-testing, and exploratory data-mining. We also identified and discussed outstanding issues we will address through additional enhancement of existing i2b2 system.

Contents

Health Information Systems

Knowledge Discovery of Biomedical Data

Exploiting HPO to Predict a Ranked List of Phenotype Categories for LiverTox Case Reports

Casey Lynnette Overby[1(✉)], Louiqa Raschid[2], and Hongfang Liu[3]

[1] Johns Hopkins University, Baltimore, USA
overby@jhu.edu
[2] University of Maryland, College Park, USA
louiqa@umiacs.umd.edu
[3] Mayo Clinic, Rochester, USA
liu.hongfang@mayo.edu

Abstract. Drug-induced liver injury (DILI) is an uncommon but important and challenging adverse drug event developed following the use of drugs, both prescription and over-the-counter. Early detection of DILI cases can greatly improve the patient care as discontinuing the offending drugs is essential for the care of DILI cases. An online resource, LiverTox, has been established to provide up-to-date, comprehensive clinical information on DILI in the form of case reports. In this study, we explored the use of the Human Phenotype Ontology (HPO) to annotate case reports with HPO terms and to predict a ranked list of phenotype categories (describing patient outcomes) that is most closely matched to the HPO annotations that are attached to the case report. The prediction performance based on our method was found to be good to excellent for 67% of case reports included in this study, i.e., the phenotype category that was assigned to the report was among the Top 3 predicted phenotype category descriptions. Future directions would be to incorporate other annotations, laboratory findings, and the exploration of other semantic-based methods for case report retrieval and ranking.

Keywords: Phenotypes · Semantic similarity · Phenotype category · Drug-induced liver injury

1 Introduction

While there remain challenges in establishing a diagnosis of drug induced liver injury (DILI) broadly, there has been an accumulation of work providing meaningful data on characteristic profiles and possible mechanisms of hepatotoxicity [1–3]. The LiverTox website[1] illustrates one effort to document standardized DILI nomenclature and methods to assess causality resulting from those efforts

---

[1] http://livertox.nlm.nih.gov.

© Springer International Publishing AG 2017
F. Wang et al. (Eds.): DMAH 2016, LNCS 10186, pp. 3–9, 2017.
DOI: 10.1007/978-3-319-57741-8_1

[4,5]. There is an opportunity to leverage this existing DILI nomenclature to better understand medications implicated in cases of hepatotoxicity. Many drugs, for example, have a characteristic clinical "signature" that can help with diagnosing conditions and assigning causality.

One promising way to mine patients' clinical signature is through electronic health records (EHRs) [6–8]. However, signs and symptoms of DILI are often captured in free-text format such as clinical discharge summaries which makes the mining task challenging. Advanced text mining approaches such as topic modeling have also been explored to generate evidence to support the causality of DILI [9]. Distinct from the previous work, we applied ontological approaches to mine a clinical signature. First, we annotate LiverTox case reports with phenotype descriptor terms from the Human Phenotype Ontology (HPO) [10–13]. LiverTox also defines phenotype category descriptions that are used to describe patient outcomes; in a further step, we associate each category description with a set of HPO terms. Finally, we predict a ranked list of phenotype categories for each case report. The ranked prediction is based on computing the semantic similarity between the corresponding sets of HPO annotations. In the following, we detail our study.

2   Materials and Methods

2.1   Resources

LiverTox provides rich and up-to-date clinical information on DILI for physicians and the public [4,5]. Among other information, it includes drug-specific hepatotoxicity reports and phenotype category[2] descriptions. Phenotype categories of DILI have been defined according to overall symptoms, signs and laboratory findings.

HPO represents human disease phenotypic abnormality knowledge with terms arranged in an acyclic graph connected by $is - a$ edges, e.g., "poor appetite" $is - a$ subclass-of "abdominal symptom".

2.2   Implementation

Other researchers have used semantic similarity to rank candidate disease diagnoses [15] and candidate genes [16] relative to selected ontology terms describing patient phenotype. Our approach is similar but ranks phenotype categories for a single disease.

We used the HPO concept recognition software [10] to annotate LiverTox phenotype descriptions and case reports. The software was used to annotate free-text in the "Clinical Summary" and "Other information" sections of phenotype descriptions. We also annotated the "Descriptions" and "Symptoms" sections of the case reports.

Three classifications of semantic similarity measures are node-based, path-based and hybrid (combining node and path-based measures). Given the graph

---

[2] http://livertox.nlm.nih.gov/Phenotypes_intro.html.

structure of HPO, we used a well studied path-based measure, the Resnik measure [14] that leverages the HPO structure to measure similarity between HPO terms. [14]. We used estimates of semantic similarity to rank DILI phenotype categories.

Specifically, Consider a phenotype category (c) annotated by the HPO term set $HPO_1 = t1_1, t1_2, \ldots, t1_m$ and case report (r) annotated by the HPO term set $HPO_2 = t2_1, t2_2, \ldots, t2_n$, as well as a similarity matrix S that contains all pairwise similarity scores of terms in HPO_1 and HPO_2. Our method computes a score for each pair of case report r and DILI phenotype category c by calculating the maximum similarity score over all pairs of HPO terms associated with r and c, respectively; see Eq. 1 [17].

$$Sim_{funSimMax}(c,r) = max \left[\frac{1}{m} \sum_{i=1}^{m} \max_{0 \leq j \leq n} s_{ij}, \frac{1}{n} \sum_{j=1}^{n} \max_{0 \leq i \leq m} s_{ij} \right] \quad (1)$$

HPOSim tools [18] were used to estimate the semantic similarity between r and c.

2.3 Evaluation

Applications of ontologies that serve a domain use case should be evaluated with respect to how well they ensure or improve the performance in particular tasks within that domain [19]. For a DILI use case, we evaluated how well our algorithm predicts the Top K phenotype categories for LiverTox case reports.

Drug-induced liver injury network (DILIN)-submitted case reports [20] were used as the benchmark. We included case reports for which a drug, according to the Roussel Uclaf Causality Assessment Method (RUCAM) scale, was "probable" or "highly probable" to have caused DILI for that case. With the assumption that case reports represent clinical diagnostic conditions for DILI, these reports provide a ground truth dataset given that phenotype category is provided for each. We use the phenotype category recorded for each case according to the R Ratio at Peak of Injury. R ratio was calculated by case report submitters by dividing the alanine aminotransferase (ALT) laboratory test measurement by the alkaline phosphatase (Alk P) measurement, using multiples of the upper limit of the normal range for both measures.

For each case report, we ranked phenotype categories by semantic similarity. Case report findings were grouped according to the position at which the true phenotype category appeared in the ranked list, i.e., *excellent* if an exact match; *good* if within the top 3; *moderate* if within the top 5; *poor* if below 5. We also compared our rankings to the ground truth phenotype categories by calculating *precision*, i.e., the fraction of the count of individual concepts with which our method found an excellent, good or moderate match, divided by the count of matches that the method identified.

3   Results

3.1   Sample of DILIN-Submitted Case Reports

LiverTox defines twelve phenotype categories of DILI[3]. There were 90 case reports submitted by DILIN that satisfied our inclusion criteria. Categories of phenotypes among those case reports were: 34 acute hepatitis, 25 mixed hepatitis, 29 cholestatic hepatitis, and 2 unrelated or not accessible. The two "unrelated or not accessible" cases were excluded from further analyses.

3.2   Phenotype Descriptors Among Phenotype Categories

Across all case reports, 79, 73, and 93 unique phenotype annotations were represented for acute hepatitis, mixed hepatitis, and cholestatic hepatitis cases. See Fig. 1 for phenotype descriptors identified among the majority of case reports grouped by phenotype category.

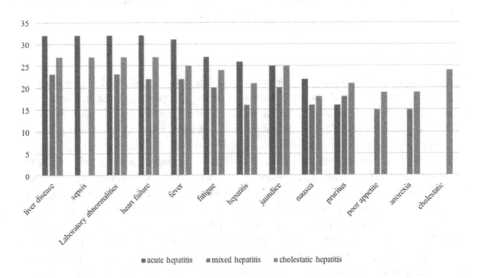

Fig. 1. The most frequent (≥15) HPO terms among case reports grouped by phenotype category

3.3   Results from Ranking Drug-Induced Liver Injury Phenotype Categories

Among the 88 remaining case reports, performance was excellent for 14(16%), good for 45(51%), moderate for 28(32%), and poor for 1(1%). Table 1 summarizes precision measures for our algorithm to rank phenotype categories for DILI cases.

---

[3] http://livertox.nlm.nih.gov/Phenotypes_intro.html.

Table 1. Precision values of our algorithm to rank phenotype categories among all, acute hepatitis (AH), mixed hepatitis (MH), and cholestatic hepatitis (CH) drug-induced liver injury cases

Match	All (N = 88)	AH (N = 34)	MH (N = 25)	CH (N = 29)
Excellent (exact)	15.9%	35.3%	0.0%	7.1%
Good (top 3)	67.0%	97.1%	61.5%	35.7%
Moderate (top 5)	98.9%	100.0%	96.2%	100.0%

4   Discussion and Conclusion

This work illustrates the use of ontological approaches to mine clinical signature in LiverTox and demonstrates the potential of using HPO to enrich LiverTox case reports for assisting the diagnosis of DILI. There are some limitations. One is the use of existing tools for ontology annotation and semantic similarity calculation which may not be optimal for our task. Additionally, our evaluation is based on case reports submitted by DILIN due to the lack of a gold standard data set.

In future work, methods used for both applications should be rigorously evaluated and compared to alternative approaches. While this work uses HPO, other ontologies can be evaluated based on how well their terms can be recognized in case reports. Rankings from calculating semantic similarity with other ontologies should also be compared with results from this work given that the method use explicit semantics in an ontology. Another area to explore are other approaches to combine term-term similarities for calculating semantic similarity. For example, here we did not incorporate the "information content" of the nodes that is based on the frequency an annotation is used. Rather we considered the presence or absence of an HPO annotation in our calculation. Hybrid measures that incorporate node-based and path-based measures such as that of Weng et al. [21] may be worth considering in future investigations.

Our findings also motivate further development of advanced approaches, such as machine learning, for phenotype category assignment for case reports. The performance of our similarity-based approach to rank phenotype categories was good to excellent for 67% of LiverTox case reports reviewed in this study. While these findings are promising, similarity measures based upon phenotype descriptors alone may not be sufficient for categorizing case reports. Linking laboratory findings, for example, may facilitate improved rankings of cases by phenotype. Moreover, characteristics of clinical course (e.g., time to onset) may be important for excluding unlikely phenotype categories. For example, the acute form of sinusoidal obstruction syndrome presents within 1 to 3 weeks of exposure to the medication and might be excluded from consideration if symptoms occur after that timeframe.

Overall, we demonstrate the use of HPO to enrich LiverTox for the assessment of patients taking medications who are suspicious of having DILI. Findings from this work motivate future approaches that incorporate other annotations

8   C.L. Overby et al.

(e.g., anatomy and clinical course), laboratory findings, as well as other methods to rank DILI phenotype and clinical outcome categories by calculating semantic similarity.

References

1. Fontana, R.J.: Pathogenesis of idiosyncratic drug-induced liver injury and clinical perspectives. Gastroenterology **146**, 914–928 (2014)
2. Lewis, J.H.: The art and science of diagnosing and managing drug-induced liver injury in 2015 and beyond. Clin. Gastroenterol Hepatol. **13**, 2173–2189 (2015)
3. Licata, A.: Adverse drug reactions and organ damage: the liver. Eur. J. Intern. Med. **28**, 9–16 (2016)
4. Hoofnagle, J.H., Serrano, J., Knoben, J.E., Navarro, V.J.: LiverTox: a website on drug-induced liver injury. Hepatology **57**, 873–874 (2013)
5. Serrano, J.: LiverTox: an online information resource and a site for case report submission on drug-induced liver injury. Clin. Liver Dis. **4**, 22–25 (2014)
6. Jinjuvadia, K., Kwan, W., Fontana, R.J.: Searching for a needle in a haystack: use of ICD-9- CM codes in drug-induced liver injury. Am. J. Gastroenterol. **102**, 2437–2443 (2007)
7. Overby, C.L., Pathak, J., Gottesman, O., Haerian, K., Perotte, A., Murphy, S., Bruce, K., Johnson, S., Talwalkar, J., Shen, Y., et al.: A collaborative approach to developing an electronic health record phenotyping algorithm for drug-induced liver injury. J. Am. Med. Inform. Assoc. **20**, e243–e252 (2013)
8. Overby, C.L., Weng, C., Haerian, K., Perotte, A., Friedman, C., Hripcsak, G.: Evaluation considerations for EHR-based phenotyping algorithms: a case study for drug-induced liver injury. AMIA J. Summits Transl. Sci. Proc. **2013**, 130–134 (2013)
9. Yu, K., Zhang, J., Chen, M., Xu, X., Suzuki, A., Ilic, K., Tong, W.: Mining hidden knowledge for drug safety assessment: topic modeling of LiverTox as a case study. BMC Bioinformatics **15**(Suppl. 17), S6 (2016)
10. Groza, T., Kohler, S., Moldenhauer, D., Vasilevsky, N., Baynam, G., Zemojtel, T., Schriml, L.M., Kibbe, W.A., Schofield, P.N., Beck, T., et al.: The human phenotype ontology: semantic unification of common and rare disease. Am. J. Hum. Genet. **97**, 111–124 (2015)
11. Kohler, S., Doelken, S.C., Mungall, C.J., Bauer, S., Firth, H.V., Bailleul-Forestier, I., Black, G.C., Brown, D.L., Brudno, M., Campbell, J., et al.: The Human Phenotype Ontology project: linking molecular biology and disease through phenotype data. Nucleic Acids Res. **42**, D966–D974 (2014)
12. Robinson, P.N., Kohler, S., Bauer, S., Seelow, D., Horn, D., Mundlos, S.: The Human Phenotype Ontology: a tool for annotating and analyzing human hereditary disease. Am. J. Hum. Genet. **83**, 610–615 (2008)
13. Robinson, P.N., Mundlos, S.: The human phenotype ontology. Clin. Genet. **77**, 525–534 (2010)
14. Resnik, P.: Using information content to evaluate semantic similarity in a taxonomy. In: Proceedings of IJCAI-1995, Montreal, Canada, pp. 448–453 (1995)
15. Kohler, S., Schulz, M.H., Krawitz, P., Bauer, S., Dolken, S., Ott, C.E., Mundlos, C., Horn, D., Mundlos, S., Robinson, P.N.: Clinical diagnostics in human genetics with semantic similarity searches in ontologies. Am. J. Hum. Genet. **85**, 457–464 (2009)

16. Masino, A.J., Dechene, E.T., Dulik, M.C., Wilkens, A., Spinner, N.B., Krantz, I.D., Pennington, J.W., Robinson, P.N., White, P.S.: Clinical phenotype-based gene prioritization: an initial study using semantic similarity and the human phenotype ontology. BMC Bioinformatics **15**, 248 (2014)
17. Schlicker, A., Domingues, F.S., Rahnenfuhrer, J., Lengauer, T.: A new measure for functional similarity of gene products based on Gene Ontology. BMC Bioinformatics **7**, 302 (2006)
18. Deng, Y., Gao, L., Wang, B., Guo, X.: HPOSim: an R package for phenotypic similarity measure and enrichment analysis based on the human phenotype ontology. PLoS One **10**, e0115692 (2015)
19. Hoehndorf, R., Dumontier, M., Gkoutos, G.V.: Towards quantitative measures in applied ontology. CoRR, abs/1202.3602 (2012)
20. Hoofnagle, J.H.: Drug-Induced Liver Injury Network (DILIN). Hepatology **40**, 773 (2004)
21. Wang, J.Z., Du, Z., Payattakool, R., Philip, S.Y., Chen, C.-F.: A new method to measure the semantic similarity of GO terms. Bioinformatics **23**, 1274–1281 (2007)

Patient Records Retrieval System for Integrated Care in Treatment of Cervical Spine Defect

Yihan Deng[1,2] and Kerstin Denecke[2(✉)]

[1] Innovation Center Computer Assisted Surgery,
University of Leipzig, Leipzig, Germany
[2] Department of Medical Informatics,
Bern University of Applied Sciences, Biel, Switzerland
{yihan.deng,kerstin.denecke}@bfh.ch

Abstract. In clinical decision making, information on the treatment of patients that show similar medical conditions and symptoms to the current case, is one of most relevant information sources to create a good, evidence-based treatment plan. However, the retrieval of similar cases is still challenging and automatic support is missing. The reasons are two-fold: First, the query formulation is difficult since multiple criteria need to be selected and specified in short query phrases. Second, the discrete storage of multimedia patient records makes the retrieval and summary of a patient history extremely difficult. In this paper, we present a retrieval system for electronic health records (EHR). More specifically, a retrieval platform for EHRs for supporting clinical decision making in treatment of cervical spine defects with the information extracted from textual data of patient records is implemented as prototype. The patient cases are classified according to cervical spine defect classes, while the classification relies upon rules obtained from the corresponding defect classification schema and guidelines. In a retrospective study, the classifier is applied to clinical documents and the classification results are evaluated.

Keywords: Information retrieval · Electronic health record · Patient cases retrieval · Cervical spine defect · Case based decision support

1 Introduction

The degenerative changes of the cervical spine are reflected in multiple aspects of observations such as area of the defect, position of the defect and additional pathology. For this reason, treatment decision making in the context of medical conditions involving the cervical spine requires evidences from all the relevant data sources (medical history, radiology, admission note) since the relevant information is distributed among these sources. The most direct way of determining and classifying these defects is to peruse the information from patient records, which is quite laborious. In clinical practice, the complete process of document selection, skimming, information extraction and aggregation is conducted manually due to missing automatic support. An efficient data collection method and

© Springer International Publishing AG 2017
F. Wang et al. (Eds.): DMAH 2016, LNCS 10186, pp. 10–25, 2017.
DOI: 10.1007/978-3-319-57741-8_2

an automatic retrieval approach are urgently required in clinical practice. An automatic classification and retrieval approach can support in:

1. Improving interoperability and increasing the efficiency of information exchange and the coverage of the defect description,
2. Increasing the accessibility of different information sources during the decision making process,
3. Facilitating the evidence based therapy and patient cohort identification.

Currently, most of the available schemas for cervical spine classification focus on the pathological changes in the spinal canal or rely upon an analysis based on the cross-sectional area [1]. They employ the investigation outcome in axial direction to grade the stages. However, the reasons of these changes and the concrete position of a lesion at the corresponding cervical vertebrae are not considered. Relying upon the graded stages only is insufficient for an effective decision making, especially in the surgical domain, since surgeons are more interested in the anatomical position and pathological changes in the target area. A more practical classification schema was required to cover the pathological changes in cervical spine. In order to provide a solution for surgical practice usage, such a defect oriented classification for the treatment of the spinal canal stenosis has been developed by Meixenberger and Leimert [2]. It is a relatively new perspective in pathological classification of cervical spine defects, since more important aspects are described in the classification schema, which can be directly applied to the further surgical decision making. More specifically, it covers the amount of the affected cervical segments, position of the defect and additional pathologies. With this schema, Daenzer et al. [3] have proposed a decision support model based on MRI image analysis. In evaluations of that model, it turned out that a classification of the defect requires sometimes additional information on the degree or location of pain and previous comparable diagnosis. Such information is captured in clinical documents of the electronic health record. However, there is no method available that supports a text-based classification of defects. In this paper, a concept for text-based defect classification will be introduced relying upon feature extraction and rule-based classification. The main challenges in this context are:

1. Gathering defect related vocabulary
2. Extracting text-based defect features
3. Classifying the defect stages using features extracted from texts
4. Indexing the corresponding aspects of defect
5. Realization of the retrieval based on extracted aspects

The classification method will be evaluated and integrated into a prototype for retrieving information on similar cases of cervical spine treatment. As is illustrated in Fig. 1, the entire system can be divided into two components: the defect classification and its knowledge base and the retrieval work flow based on classification knowledge. The two components will be described in the following sections separately.

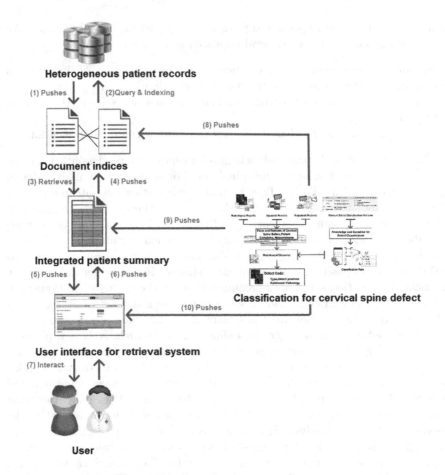

Fig. 1. Flowchart for the retrieval system

2   Knowledge Base for Cervical Spine Classification

2.1   Defect-Based Classification Schema

In this section, we are describing the classification schema underlying our rule-based approach [2]. As an interdisciplinary approach, the schema depicted in Fig. 3 was developed by neurosurgeons and radiologists to assess the disorders and pathologies in cervical spine. The experts from anatomy and medical informatics have also provided their suggestions for the definition of defect grading. In comparison to other schemas, the focus of this new grading method lies on the anatomic landmarks and direct topographical descriptions. It provides a straight access to the actual pathology. The classification schema considers three aspects: type, defect position and grade of additional pathology. The type refers to the number of affected segments and their relations to one another, for example, single, several subsequent or skip lesions. According to different clinical findings

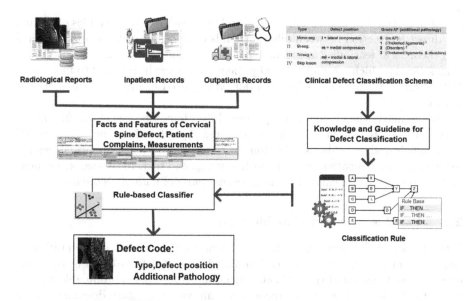

Fig. 2. Schematic graph of the defect classification process

depending on the position of intraspinal pathology, the second point of differentiation is the defect position, i.e. lateral or medial affection of nerval tissue. These two positions may also co-occur at the same time.

The third characteristic in the schema pays special attention to anatomic and functional disorders that have not been taken into account in the existing classifications: besides compression caused by prolapse of the nucleus pulposus and osseous humps, thickening of ligaments like the yellow ligaments and the posterior longitudinal ligament as well as structural disorders (listhesis, kyphosis, hyper lordosis or rotation) may also affect viability in a surgical relevant way. This defect classification schema enables a very detailed description of a defect by using only three characters, for example bi-m2 referring to a bi-segmental stenosis with medial compression of the spinal cord with additional listhesis.

As there are several classification schemas regarding different clinical usage, a brief comparison can point out the differences and advantages of the defect classification schema used in this work (see Fig. 3). For comparison, we consider the classification system based on kinematic magnetic resonance imaging in cervical spondylotic myelopathy [1] and the qualitative grading of severity of lumbar spinal stenosis based on the morphology of the dural sac on magnetic resonance images [4].

In comparison with the aforementioned schema, our defect classification schema is specially designed for surgical practice. It covers more defect information and categorizes additional pathological information. Besides these advantages, a much more direct description of the locoregionary and anatomical causes facilitates the planning and the feasibility of surgical treatment, whereas focusing on consequences like impression of the dural sac leaves several variables regarding the direct interaction between different kinds of tissues. In addition to technical and theoretical feasibility, a precursor of this schema already underwent practical testing, in which ten anonymized patient cases were introduced to more than 100 experienced surgeons, who in majority approved the practicability of that schema.

In summary, the classification proposed by Meixensberger et al. [2] enables direct access to surgical decision/planning by stressing the anatomical/ morphological abnormalities. Mentions or occurrences of such abnormalities are easily detectable in patient records both in image and text. In this work, we focus on their detection and classification in textual parts of patient records.

2.2 Text-Based Defect Classification

Hitherto, surgeons have based their decisions on the manual interpretation of radiology images showing the location and degree of defects (see Fig. 3). However, studies have found that this classification is often not reproducible and far from standardized. The patient records contain supplementary documents with additional important evidence which is currently not considered or available at the time of decision-making. Nevertheless, the defect situation and patient status described in the radiology report, anamnesis and admission note are crucial for classification and follow-up therapy planning. To improve the existing defect classification process, we have augmented the image-based classification with an approach relying on information extraction for defect classification from the clinical narratives. The following steps need to be performed:

1. The relevant pieces of information need to be identified in the text.
2. Contexts in which information occurs need to be determined.
3. Rules for mapping extracted information into a defect category mapping need to be defined and the automatic categorization needs to be realized.

More specifically, the classification task is as follows: For each of the three aspects of the underlying classification schema (type, defect position, additional pathology), one out of three or four category needs to be assigned (see classification schema for the number of categories). Only documents with mentions of all three defect-related features will be considered in our pipeline. Defect types, defect position, and additional pathology are identified and classified by the extraction pipeline and by applying classification rules.

We are addressing these issues by developing the following components (see Fig. 2): First, we establish a defect terminology. The defect-related terms are initially gathered from clinical narratives. They are manually classified into categories such as disease, or anatomical concepts at the area of cervical spine. Second, the extraction pipeline is configured. The concepts relevant for defect classification are extracted by a concept mapper that identifies mentions of defect terms. Terms that describe diagnoses and anatomical concepts are mapped to concepts of corresponding standardized medical terminologies and classification

Type		Defect position	Grade AP (additional pathology)
I	Mono-seg.	l = lateral compression	**0** (no AP)
II	Bi-seg.	m = medial compression	**1** (Thickened ligaments) [1]
			2 (Disorders) [2]
III	Tri-seg.+	ml = medial & lateral	**3** (Thickened ligaments & disorders)
IV	Skip lesion	compression	

Fig. 3. Defect classification schema

systems. Then, regular expressions are used to detect acronyms, conventional expressions and special word combinations.

Third, classification rules are defined. The rules regulate the mapping between extracted features and defect categories. The aforementioned defect classification schema is transformed into corresponding classification rules considering the three main types: defect type (the amount of defect segments), position (medial, lateral or medial lateral) and additional pathology (thickened ligaments, disorders). In the final system, a clinical data interface connected with the hospital information system (HIS) and Picture Archiving and Communication System (PACS) needs to be designed to get the automatic import of patient records while employing the corresponding authentication and authorization process to protect patient data privacy. In this paper, we will focus on the rule-based classification using textual features and the evaluation of extraction results. The performances of these two steps will be evaluated. In the following, the single steps and components are described in more detail.

2.3 Defect Terminology

For detecting defect specific features in clinical texts, a list comprising defect terms in German was generated. At a first step, a clinical expert has annotated the defect relevant terms in patient records. Then, the terms were categorized into several subcategories such as anatomy, symptom, pathology and positions. Meanwhile, the relevant terms in the terminology list were directly linked to the concepts of existing terminologies that are Radlex (German), ICD 10 (German) and MeSH 2010 (German) as expansion to increase the recognition rate of the concept mapping. In total, 311 defect specific terms were included into the concept dictionary.

2.4 Extraction Pipeline

The extraction pipeline is established based on an adapted version of UIMA (https://uima.apache.org/) provided by Averbis GmbH (https://averbis.com/). Two types of concept mapping are implemented in the framework: the exact mapping and the segment mapping. They have been configured to our task by including the cervical spine vocabulary (defect terminology, see section above).

Exact mapping searches only for fully matched terms in the corresponding semantic scope, whereas the segment mapping performs a fuzzy matching considering morphological variations of terms. Further, the contiguous match strategy is employed to obtain the longest match of contiguous tokens within a sentence. Besides, a German negation list is additionally exploited by the concept mapper. Three main types of negations in German were defined, namely post negation (nicht vorhanden (non-existent)), pre negation (frei (free)) and pseudo negation (nicht sicher, ob (not sure whether)).

Moreover, regular expressions were defined to recognize dates, measurements and doses expressed in text. Addressing domain- or task-specific vocabulary, the concept mapper was complemented by additional regular expressions, for example to identify location descriptions expressed as coordination structures. Consider the following example text *HWK 3 und 4* (vertebral body 3 and 4 of cervical spine). The coordination structure needs to be recognized by a regular expression. The first part of the phrase, *HWK-3*, has been captured by the concept mapper, the second part *und 4* needs to be recognized by regular expressions and transformed into standard form of *hwk3/4*. In summary, the anatomical concepts regarding intervertebral disc and positions as well as the additional pathologies are extracted through the aforementioned pipeline.

2.5   Classification Rules Definition

Classification rules are developed to formalize the defect classification schema. The efficiency and effectiveness are main concerns for rule formalization [5]. Therefore, we choose a declarative way to model a rule, which can avoid the complexity of procedure description. Additionally, an easy modification and high reusability are important for the maintenance of knowledge rules. Several methods for rule formalization are available [6]. Among them, the logic programming language Prolog is normally used to transform the knowledge rules. However, our extraction pipeline is implemented in Java. The rules are formalized and hard-coded in Java program code. According to the defect classification schema, three levels of knowledge can be defined. We are organizing it from generic to specific:

1. Common sense includes the temporal comparison, input and output logic.
2. General medical domain knowledge such as anatomical structure and position.
3. Defect specific knowledge indicates the category specification based on the classification schema and the implicit connections between symptoms and categories according to practical surgical experience.

Since the classification foresees only a limited number of defect types, the possible lexical variations of describing defect types are listed explicitly in the rules. For example, a fact can be defined with Java like if(defect("hwk3/4") == true&&no negation) defecttype++;. The input is the normalized form of intervertebral disc, which is mapped with the amount of defect segments. 15 linguistic variations of description of defect types and defect combinations from mono to tri+ have been summarized in the rules. Besides, the incontinuous serials of input will therefore be classified into type skip. In this classification schema, defect stands for the spinal canal stenosis, i.e. low degree of compression will not be classified as defect.

The position should be handled in two types: The first type is the explicit description. It can be directly mapped with position determination functions. For deciding of defect at both medial and lateral position, we have added several specific rules with clinical salience. In addition, the anantomical concept has provided implicit hint for the position. For example, Facettengelenk (intervebral joint or facet joint), uncovertebral (uncovertebral joints), foramina (intervertebral foramina), Nervenwurzel (spinal nerve root) indicate also the defect at the position of lateral.

The additional pathology contains four stages: namely, no additional pathology (0), thickened ligaments (1), disorders (2) or both thickended ligaments and disorders (3). Disorders represent listhesis, rotation, kyphosis, hyper lordosis and steep. Since these concepts are distinct and unique, they are simply summarized in the knowledge base and provide the mapping between extracted features and grading. For example, if (pathologyCheck(object).equals(Ligamental flava) pathology = '1' has defined the pathology in Ligamenta flava as grading one.

3 Retrieval Architecture

3.1 Patient Record Indexing

Based on the automatic extraction and rule-based classification for the cervical spine defect, the important aspects about the defect pathology has been extracted. For indexing the patient records, we are using the method of inverted indices which is a relatively mature technologies in the field of information retrieval [7]. As is illustrated in Fig. 4, the weighting value is:

$$W_{i,d} = tf_{i,d} * log(n/df_i)$$

where

$$d = tf_{i,d} * log(n/df_i)$$
$$tf_{i,d} = frequency\, of\, term\, i\, in\, document\, j$$
$$n = total\, number\, of\, documents$$
$$df_i = the\, number\, of\, documents\, that\, contain\, term\, i$$

The ranking value of one patient record is the sum of all the terms weighting values. The high ranking value indicates the high relevance. For example (see *example* in Fig. 4), if one keyword has a term frequency of 2, an inverse document frequency of 1.9 (10 appearances in 800 documents) a, the W value is 3.8 (achieved by applying the formula). The sum value in Fig. 4(a) example is the sum of all W values of queried terms in one document, the large sum value indicates the high similarity between the query and documents. Besides the sum value, similarity measures such as cosine similarity or structural similarity are also be applied to obtain the relevant electronic health records.

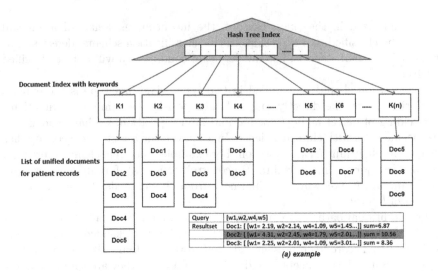

Fig. 4. Index structure of the retrieval system

3.2   Retrieval Interface

The ambition of the retrieval interface is to provide an integrated access to the defect relevant patient information in the treatment of cervical spine stenosis, which bridges the heterogeneous clinical information system and fill the gap of data storage in the clinical work flow. Physicians can therefore use the retrieval interface to find the desired patient record in all different clinical information systems. Meanwhile, the corresponding patient cohort for the cervical spine and the quality studies as well as the communication between attended care providers can be eased. Additionally, the similarity between different patients can be compared not only at term level but also through the metrics derived from the defect classification schema. The current retrieval system can present patient records regarding extracted aspects and also provide the full text search. As it can be seen in Fig. 5, five aspects are shown in the retrieval interface to support the user in navigating through the patient records. The first three aspects shown in the navigation bar on the left are the three dimensions from the classification schema (defect type, position and additional pathology). The other two search aspects are the defect relevant terms and ICD-10 codes. Physicians can get a fast access to any subcategory of the these five aspects. Besides this faceted search, the clinical user can also search for patient records using the query input field. Besides the full text search in the search field, the user can also constitute the query through the query builder, see Fig. 6. The query referring to corresponding aspects can be combined together. With the application of query builder, the desired query can be saved as default search setting.

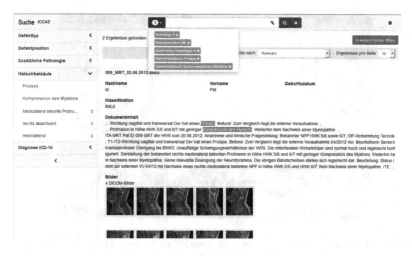

Fig. 5. Screenshot of the clinical retrieval platform: *The platform combines the aspect search (left) and full text search field (middle search field) and patient comparison (blue button). The paired patient record with textual records and images is presented in below. The patient information and defect codes are listed in the middle.* (Color figure online)

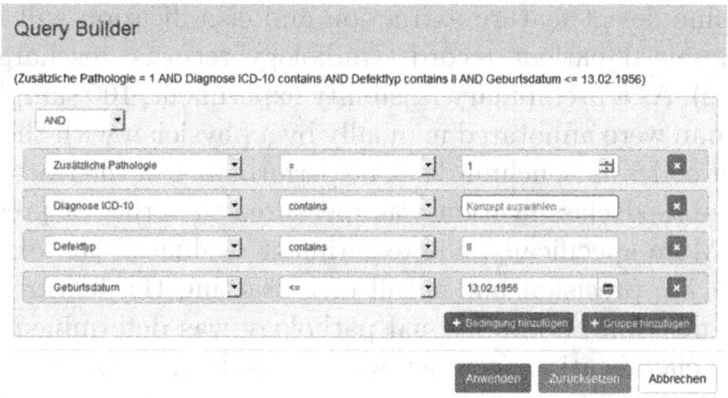

Fig. 6. Query builder of the retrieval platform: *The relevant aspects can be connected with predefined value through Boolean operators, For example, the ICD 10 terminology (stenose) and defect type (mono defect) as well as the additional pathology (rotation) can be saved as default search setting*

The textual reports and the corresponding images are presented in text and image pairs through the interface. Physicians will also have the possibility to send their feedback to the system regarding the classification (see Fig. 7), i.e. they can hand-code the defect category if the suggested automatic one seems to be incorrect. The feedback will be stored as additional classification labels together with patient records and can be used later on in adapting the classification rules.

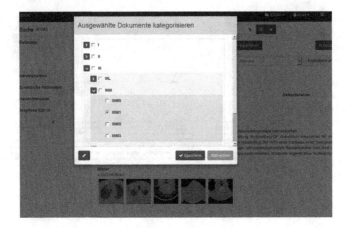

Fig. 7. User Interaction Box. *A manually updated classification of the defect category can be added through the cascading radio buttons by the clinical user.*

4   Evaluation

The cervical spine defect feature extraction and classification will be evaluated based on anonymized patient records (radiology records, discharge summary, outpatient cards). As a preliminary feasibility experiment, 100 samples of patient records in German were annotated manually by a physician with their defect features and defect codes as benchmarks. The performance of the feature extraction and of the rule-based classification was measured in terms of precision, recall and accuracy. More specifically, the experiment evaluated the extraction part of the pipeline, i.e. precision and recall of extracting the features on cervical spine segment, position and additional pathology was determined. In a second experiment, the classification accuracy was assessed. The quality of the retrieval remained so far unconsidered.

5   Results

5.1   Performance of the Extraction

The annotated corpus comprises 100 anonymized patient records. Four records have been detected as unusable for the text-based classification due to the incomplete description of the three required classification categories. For example, the main finding is related to intramedullary spinal cord tumor while the spinal canal stenosis has only been briefly mentioned with few words as secondary finding.

It can be seen in Table 1, that the precision of the extraction for the three feature categories already achieves optimal results. However, the recall needs still to be improved. The detection of cervical spine segments has the highest miss rate due to the non-uniformed notation and abbreviations of these concepts.

Table 1. Extraction results

Extracted element	Precision	Recall	F1
Cervical spine segment	100%	85%	91%
Position	100%	99%	99%
Additional pathology	100%	97%	98%

Table 2. Classification results

Defect category	Accuracy
Type Mono	90%
Typo Bi	95%
Type tri+	94%
Position medial (m)	91%
Position lateral (l)	98%
Position m & l	94%
AP 0	99%
AP 1	97%
AP 2	100%
AP 3	98%

5.2 Performance of the Rule-Based Classification

With the same annotated corpus, the performance of the rule based classification is evaluated. The classification accuracies were calculated for each classification category separately, since all types are assumed to be independent events. No defect falling into the category skip was contained in our data set, hence, in this experiment, we will firstly focus on the evaluation of the three common types of defects (mono, bi and tri+). The accuracy is the proportion of true results in each category. As is illustrated in Table 2, the classification of the feature "additional pathology" has generally a high accuracy, while the classification of type (bi) has shown better performance than the classification of the other two types (mono and tri+).

In summary, the extraction pipeline provides defect features with high precision. At first, the classification of position resulted in a relatively high error rate by the first rule definition and the classification of type (number of defect) has shown deviations in various degrees. After several rule extension and updating, new term features have been considered and added to the rule base. The classification has in this way reached a clearly better accuracy. The detailed analysis and possible improvements will be discussed in an error analysis in the next section.

6 Discussion

6.1 Error Analysis

The feasibility study showed that the current extraction method based on concept mapping and regular expressions achieves optimal precision and good recall. The recall can be improved through the extension of the terminology dictionary and additional regular expressions. Based on the current knowledge base and rules, the explicit description for type and position can be detected with good performance. However, the implicit position reflected in anatomical concepts and different severities of protrusion at cervical spine segments still needs to be extracted for a better category determination. Moreover, modifiers around the mentions of the defect segment in one semantic scope (sentence) should be considered besides negation, since the defect segment and its position are closely related to each other, whereas the analysis at document level cannot guarantee an accurate mapping between a subject and its modifiers.

The main source or extraction errors lies mainly in the identification of the cervical spine segment. After manually inspection, it became clear that a varying vocabulary leads to a high miss rate. The regular expressions regarding three variations defined in our extraction pipeline are clearly insufficient to deal with the various notations used by different physicians. As one possible improvement, possible terms and abbreviations will be collected from the current corpus and will be summarized in regular expressions to increase the recall. Additionally, basic natural language processing methods could be applied to abstract from lexical variations.

According to the current knowledge obtained from guidelines, only the explicit description of medial position using adjectives is mapped, while the position implied by anatomical concept has only partly been summarized. Furthermore, the writing style and conventions of the defect position show a large lingustic variety. Therefore, the rule description of medial position needs to be extended under the help of our clinical experts. Referring to the classification error by the recognition of type (number of defect segments), the reasons are twofold: first, disc protrusion may not always lead to a spinal canal stenosis. If the segment has only been described as protrusion, further information related to severity and the final judgment of this symptom should be extracted to decide whether it is really a spinal canal stenosis. Second, the semantic scope for negations in German is obviously larger than in texts written in English. Especially in clinical narratives, a larger context length need to be considered to determine the negation. Plenty of false positives by type classification in our experiment were caused by an incorrect extraction of negations in those complex sentences with several clauses.

6.2 Limitation of the Study and Approach

The limitation of the experiment is its small data amount and simple evaluation method. More specifically, the correlation between the three defect categories has not been evaluated. For example, it would be interesting to see the co-occurrence between additional pathology and defect numbers, which can also facilitate the generation of rules for defect number determination. In addition, the utilities of rules and length of the decision paths for each classification can be evaluated to test the efficiency of rules, so that the irrelative and inefficient rules can be eliminated. In addition, for our experiment, we considered texts from one department of one hospital. Through inspections of texts of other clinics, we learned already, that terminology usage differs. Thus, adaption needs to be conducted either in rules or in the terminology list to reduce the errors due to linguistic variations. The hard coding of rules has the benefit of a fast calculation and a wrapper for interpreting a rule base is unnecessary. Another possibility would have been to create an external rule base that could be more easily adapted also by layman. Given the limited number of rules and the Java infrastructure that we used (Averbis pipeline), we decided for the hard coding of classification rules.

Besides a rule-based method, machine-learning is a frequently used method for classification tasks. For example, Zhou et al. applied a dynamic language model and a Naïve Bayesian classifier to classify radiology records based on expert annotation [8]. Claster et al. employed the self organizing map (SOM) to learn the correlation between clinical events regarding overuse of radiological diagnosis for children in an unsupervised manner [9]. However, applying machine learning requires a comprehensive annotated data set for learning classification rules or classifier respectively, which was unavailable for the specific use case we were considering. Beyond, automatically trained rules are often not understandable or reproducible by humans. In contrast, our rule base has the benefit of making these rules easy adaptable and re-traceable.

6.3 Extension of Ranking Method

The retrieval platform based on keyword ranking and hash tree has been implemented through Apache Solr framework[1]. The system has received a first clinical evaluation in practice. A physician from an inpatient department has used the system to reduce the treatment duration, while the surgeons in the neurosurgical department applied the system to do the retrieval of the similar cases and planed for the surgical operation. They confirmed the usefulness of the system. However, we still need to perform a structured evaluation of the entire retrieval system and its usability. Further, a more treatment oriented ranking method of the retrieval results need to be developed.

A ranking method underlying that bases on clinically relevant aspects is still not well studied. A possible approach for ranking would be by employing the prior probability so that the weighting considering the classification schema can be used to adjust the ranking schema. More specifically, the ranking formulation can be extended with the probability P from classification schema, which indicates the user preferences.

---

[1] http://lucene.apache.org/solr/.

$$W_{i,d} = tf_{i,d} * log(n/df_i) * P_d$$

where

$$P = defect\,weight\,for\,document\,d$$

so that the defect classification and user predefined profile can be used to represent the severity of the patient whereas the patient records with significant defect and desired pathology will be ranked higher than other general defect combinations.

7 Conclusion

Due to the frequent occurrences of pathological changes of the cervical spine as well as the numerous possibilities of degenerative changes, the clinical demand for an easy to use and reliable tool for direct classification and for retrieving documents referring to pathological changes of various origins is obvious and urgent. We suggested an approach using information extraction and rules for defect classification. As a pilot study, preliminary experiments have been conducted with the support of surgeons. The usefulness of the automatic classification was confirmed by clinical experts. Our study revealed several concrete challenges of the textual based classification: (1) The diversity of notation, abbreviation and writing style is the main obstacle of an automatic defect classification. Improved mapping methods should be applied to close the gap between expressions in clinical narratives and in standard knowledge bases. (2) The interpretation of implicit features from clinical narratives needs to be considered, e.g. anatomical concept that implies the position. More empirical knowledge should be summarized to overcome the difficulties by the mapping between implicit features and standardized classification rules. (3) The utility of context information is still not well exploited. More contexts should be considered to reduce the ambiguities by the recognition of the classification condition, e.g. the certainty of the physician, the severity of the protrusion should also be considered.

In future, a fine-grained hierarchy will be created for the current terminology dictionary instead of the flat structure, so that the semantic distance between different terms can be compared. More features such as severity, certainty are planned to be extracted. The analysis will be conducted at sentence level instead of document level to increase the mapping rate between subjects and modifiers. Further, an additional clinical study will be organized for collecting more rule definitions from empirical evidence. The final objective of the automatic classification is to determine therapy recommendations. For this purpose, for each defect class and associated features, therapy options need to be made available. Then, the system will be able to provide a therapy recommendation and undermine the decision with relevant cases of previously treated patients.

References

1. Muhle, C., Metzner, J., Weinert, D., Falliner, A., Brinkmann, G., Mehdorn, M.H., Heller, M., Resnick, D.: Classification system based on kinematic MR imaging in cervical spondylitic myelopathy. Am. J. Neuroradiol. **19**(9), 1763–1771 (1998)
2. von Sachsen S.: Computer aided defect classification for model-based therapy of cervical spinal stenosis. In: International Conference on Biomedical Engineering and Systems (ICBES), Prague, Czech Republic (2014)
3. Daenzer, S., Freitag, S., von Sachsen, S., Steinke, H., Groll, M., Meixensberger, J., Leimert, M.: Volhog: a volumetric object recognition approach based on bivariate histograms of oriented gradients for vertebra detection in cervical spine MRI. Med. Phys. **41**(8), 082305 (2014)
4. Schizas, C., Theumann, N., Burn, A., Tansey, R., Wardlaw, D., Smith, FW., Kulik, G.: Qualitative grading of severity of lumbar spinal stenosis based on the morphology of the dural sac on magnetic resonance images. Spine (Phila Pa 1976) **35**(21, 9), 1919–1924 (2010)
5. Samwald, M., Fehre, K., de Bruin, J., Adlassnig, K.-P.: The arden syntax standard for clinical decision support. J. Biomed. Inform. **45**(4), 711–718 (2012)
6. Huang, Z., Teije, A., Harmelen, F.: Rule-based formalization of eligibility criteria for clinical trials. In: Peek, N., Marín Morales, R., Peleg, M. (eds.) AIME 2013. LNCS (LNAI), vol. 7885, pp. 38–47. Springer, Heidelberg (2013). doi:10.1007/978-3-642-38326-7_7
7. Kelly, D., Sugimoto, C.R.: A systematic review of interactive information retrieval evaluation studies, 1967–2006. J. Am. Soc. Inform. Sci. Technol. **64**(4), 745–770 (2013)
8. Zhou, Y., Amundson, P.K., Fang, Y., Kessler, M.M., Benzinger, T.L.S., Wippold, F.J.: Automated classification of radiology reports to facilitate retrospective study in radiology. J. Digital Imaging **27**(6), 730–736 (2014)
9. Claster, W., Shanmuganathan, S., Ghotbi, N.: Text mining in radiological data records: an unsupervised neural network approach. In: First Asia International Conference on Modelling Simulation, AMS 2007, pp. 329–333, March 2007

Managing, Querying and Processing of Medical Image Data

IEVQ: An Iterative Example-Based Visual Query for Pathology Database

Cong Xie[1], Wen Zhong[1], Jun Kong[2], Wei Xu[3], Klaus Mueller[1],
and Fusheng Wang[1(✉)]

[1] Department of Computer Science, Stony Brook University, Stony Brook, USA
{coxie,wezzhong,mueller,fushwang}@cs.stonybrook.edu
[2] Department of Biomedical Informatics, Emory University, Atlanta, USA
jun.kong@emory.edu
[3] Brookhaven National Lab, Computational Science Initiative, Brookhaven, USA
xuw@bnl.gov

Abstract. Microscopic image analysis of nuclei in pathology images generates tremendous amount of spatially derived data to support biomedical research and potential diagnosis. Such spatial data can be managed by traditional SQL based spatial databases and queried by SQL for spatial relationships. However, traditional spatial databases are designed for structured data with limited expressibility, which is difficult to support queries for complex visual patterns. Moreover, SQL based queries are not intuitive for biomedical researchers or pathologists.

In this paper, we investigate the expressive power of visual query for spatial databases and propose an effective yet general Iterative Example-based Visual Query (IEVQuery) framework to query shapes and distributions. More specifically, we extract features from nuclei in pathology databases, such as shape polygon nuclei density distribution, and nuclei growth directions to build search indexes. The user employs visual interactions such as sketching to input queries for interesting patterns. Meanwhile, the user is allowed to iteratively create queries, which are based on previous search results, to finely tune the features more accurately to find preferred results. We build a system to enable users to specify sketch based queries interactively for (1) nuclei shapes, (2) nuclei densities, and (3) nuclei growth directions. To validate our methods, we take a pathology database [11] consisting of hundreds of millions of nuclei, and enable the user to search in the database to find most matching results through our system.

Keywords: Pathology data · Visual query · Visual analysis

1 Introduction

Digital pathology images are generated through scanning human tissue specimens with high resolution microscope scanners. Examination of high resolution images enables more effective diagnosis, prognosis and prediction of cancer and

F. Wang et al. (Eds.): DMAH 2016, LNCS 10186, pp. 29–42, 2017.
DOI: 10.1007/978-3-319-57741-8_3

other complex diseases. Pathology image analysis segments large amounts of spatial objects, such as nuclei and blood vessels, along with many other image features. Such spatially derived features are used in many analytical queries [13] to support biomedical explorations. The pathologists usually want to finding potential cancer regions by examing important properties of nuclei in large amount of pathology images, which include:

- **Shapes of cells.** Misshapen cell nuclei are frequently observed in malignant tumor and other diseases.
- **Distribution density of cells.** There can be a rapid density change of cell distribution from a normal region to a potential tumor region.
- **Growth directions of cells.** In a potential tumor region, the cells can be distributed around an empty area. All the directions of nuclei are pointing outside (the direction of major axis). This happens because cells inside are dead due to lacking nutrition, and all the other cells expand outward (Fig. 1).

Traditional SQL based spatial databases support structural queries, such as nearest neighbor search, containment, and window search. However, it is very difficult to express queries to find cells with complex patterns using standard SQL-based spatial queries for pathology databases, because of the following challenges:

1. Some features are hard to be translated into queries manually. For example, the distribution of cells in an interesting region is too complex to be defined using query language. If the raw data, images or visualizations can be used as queries, it will be more friendly to the user.
2. When the features of the desired results are in different granularities. For example, cells in a window can be distributed uniformly in general, while the density in the center of the window can be much lower than other parts. The user needs to use a top-down method: search windows with main desired features first (e.g., query cells with uniform distribution in general), then refine result with detailed features (e.g., adjust the distribution in some local area). Therefore, an iterative query process will be helpful for the users.
3. In most of the times, it is much easier for an expert to answer "yes" or "no" than determine the details of desired features. In other words, if the query system can provide potential candidates, or exemplars from the dataset, it is straightforward for the user to judge whether the candidates are desired or not. For example, the user wants to search a window with specific distribution but he or she doesn't remember the detailed distribution in the corner. When given a distribution from the result set, the user can tell whether they are preferred. In this way, the user can be guided step by step to approach the most preferred results with the exemplars provided by the database.

Based on these factors, we propose an iterative visual query process instead of traditional SQL query statement. Query is created by inputting data or visualizations, such as drawing the sketch of the target. For a non-database expert, this is much easier than defining a complex model or SQL query statement.

During the query process, the user is allowed to add or modify the features of the input query incrementally to find preferred results gradually. This top-down approach first searches for the most general information and then reveal the detailed patterns. During each step of the iterative exploration, if an intermediate search result is preferred, the user can refine the query based on the current result. This is an interactive influence process and it terminates until the user finds the desired and satisfied result. On the one hand, since the intermediate results in every iterative step are shown as user's feedback, the user can lead the search process in right direction. On the other hand, the results guide the user's decision and provide opportunities in search of the best final results.

As far as we know, this is the first work focusing on visual query for pathology database. The rest of the paper is structured as follows: Sect. 2 reviews related works. The pathology dataset and the general frameworks of our method are described in Sect. 3. Sections 4, 5 and 6 propose the query approaches of nuclei shape, nuclei density and nuclei growth direction respectively. Finally, the paper is concluded in Sect. 7 with a brief discussion of limitations and future directions.

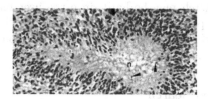

Fig. 1. An example of potential tumor region from the pathology dataset. There is an empty area inside, and all the growth direction of cells are outwards.

2 Related Work

2.1 Spatial Query

Spatial query is often used in geographical dataset. Some existing works dealt with the visual query of the vehicle trajectory [3,5]. De Silva et al. [4] proposed a system for retrieving human trajectory patterns from tracking data captured within a large geographical area, over a long period of time. David et al. [3] combined spatial querying and mobile technologies to formulate queries by drawing the desired spatial configuration on touch-sensitive screens, avoiding complex statements in some SQL-like query language.

In recent years, the fast development of computer in medical areas provided doctors opportunities to use database for pathology images query. Spatial database techniques are usually applied to find interesting cells in pathology dataset [13,14]. For example, the following simple query tasks: finding the nearest blood vessel of a cell, computing the variation of intensity of each biological property associated with the cell in respect to the distance, and returning the density distribution of blood vessels around each cell, can be solved by basic spatial queries such as: finding the location of a cell, finding cells inside a 2D window and etc. However, most of the existing spatial queries cannot be able to deal with more complicated queries such as searching the density of nuclei, because these kinds of queries are hard to be translated into a spatial query statement.

2.2   Sketch-Based Search

Gesture DB [7] allowed some low-level operation of queries, while for complex tasks, sketch-based query can be employed. For example, in large image database, image could be retrieved by matching sketched templates over the shapes [2,6]. QuerySketch [15] queried a set of time series of stock price information using a sketching interface. Wei et al. [16] searched streamlines in a vector field from input sketch to find interesting patterns in the vector field. Average Explorer [18] allowed the user to sketch the desired features of image to retrieve the results which matched the input in an iterative process.

Our approach uses visualizations such as sketch input to replace SQL queries. However, creating a sketch may be hard since the user may be unclear about the details of preferred patterns. Because making a choice is easier than defining the features, we allow the user to create the input based on exemplar data from the intermediate results.

2.3   Human-Assisted Search

There are many works focusing on human-assisted search. Yang et al. [17] explored large dataset with human interactions. Parameswaran et al. [9] dealt with the problem of human-assisted graph search. They proposed a way to guide the user to approach the goal by asking questions that if the current result is preferred or not. Anish et al. [10] researched on finding the most succinct and accurate view definition given a database instance and a corresponding view instance. Besides, some other works combined human assistance as query feedback to direct the user's exploration. Behrisch et al. [1] let the user select interesting features of multidimensional data during exploration, to reduce the preferred data space iteratively.

However, most of the above works only provided feedback framework by simply answering "yes" or "no" without modifying the user's customized feature. The search process can be inefficient if the system does not find the best question to ask. Our method enables the user to response to the system with interesting features directly by sketching interesting instances through visual interactions.

3   Dataset and Approach Overview

3.1   Pathology Dataset

We adopt the dataset from Pathology Analytical Imaging Standards (PAIS) [11, 12] data management system, which employs a spatial DBMS based implementation for data management. Containing sets of slides of pathology images, it provides information about morphological and functional character- istics of biological systems. Following border extraction techniques, automatic algorithms are used to identify the cell nuclei polygons in images. Other features are extracted from the polygon of nuclei, such as position, area, perimeter of the nucleus polygon.

Based on the extracted shape polygons of nuclei, as mentioned in Sect. 1, our approach mainly focus on finding the nuclei polygons of specific shape, finding windows of image with interesting density distributions, and finding the windows of image with interesting growth directions of cells.

3.2 Approach Overview

Figure 2 shows the overview of our approach for visual query of the cell nuclei in pathology dataset. Firstly, the query input is in the form of visualization (e.g., hand sketching) (Fig. 2(a)). Then the SQL query which matches the visualiza- tions best is found (Fig. 2(b)). The database returns the results of the SQL and visualizes them (Fig. 2(c)). The user is allowed to select the most desired data from the results (Fig. 2(d)). The intermediate results can also be used and mod- ified as new input query for next search iteration. The process is iterative until the most preferred result is reached (Fig. 2(e)). As an example of shape query to illustrate the approach in Sect. 4.

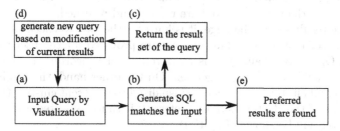

Fig. 2. The pipeline we proposed which guides the user to query their desired data in a visual analytics approach.

4   Shape Query of Nuclei

Shape query is proposed for detecting abnormal cell nuclei with special shapes which are interesting to the user.

4.1   Sketch Query

The user is allowed to draw the desired shape on canvas as input of the query (Fig. 3(a)). To help the user tunes their sketch input finely, 30 control points around the contour will be sampled uniformly (Fig. 3(b)). The user can drag the

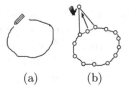

(a)   (b)

Fig. 3. (a) Query input by sketching the shape. (b) Modify the shape sketch by dragging the control points.

control points to modify the shape. This sketch will be translated into SQL and submitted to the database.

We adopt turning function [8] for shape similarity measure. For a polygon, it scans all the edges in turns, and calculates an array of rotation angle between each edge and its previous edge. The array is used as turning function for similarity comparison. Since contours of objects in digital images are distorted due to digitization noise and segmentation errors, it is desirable to neglect the distortions while, at the same time, preserve the general perceptual appearance sufficient for object recognition.

All the turning functions of nuclei polygons are pre-computed. For the user's sketch polygon input, we calculate the turning function of control points and match it with every nucleus in the database based on Euclidean distance measure. The user is allowed to set a distance threshold, and the nuclei with a distance smaller than the threshold are ranked and returned.

Since the result set can be huge, it is hard for the user to check each nucleus in the result set one by one. Two options are provided for the user to explore the results. On the one hand, the user can rank nuclei according to the distance between the sketch input and target cell. On the other hand, the result set can be clustered and cluster centers can be visualized to represent clusters (Fig. 4(b)) as well. If one of the clusters is more preferred, it can be selected and the nuclei in that cluster will be shown (Fig. 4(c)).

4.2 Iterative Search

If the user find any preferred shape from the result (Fig. 4(c)), it can be used as input sketch for next query. If the selected shape needs to be modified, the user can drag the control point of the polygon of the shape to adjust it (Fig. 3(b)). For example, the user wants to find cells with desired nuclei shape in a pathology dataset. Since the user only has a general idea about the patterns of shape without detailed features in the beginning, the examples from intermediate query results are essential for the user to select preferred shapes and results.

– **Step 1:** The user input a visualization of shape s_0 as input (Fig. 5(a)). In this example, the input is the desired sketch of shape by hand. s_0 describes the desired shape pattern in general. However, it may not match exactly. The user is allowed to add or modify features to it in the following process.

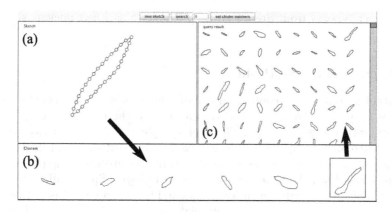

Fig. 4. The user interface of shape query of nuclei in pathology dataset. (a) The user sketches the desired shape on the canvas. 30 control points are added in the contour for fine tuning of the sketch. (b) The query result set are clustered into 6 clusters. The center of each cluster is shown to the user. The cluster number can be set by the user in the top menu. The set of cell nuclei in the last cluster is selected and shown in (c).

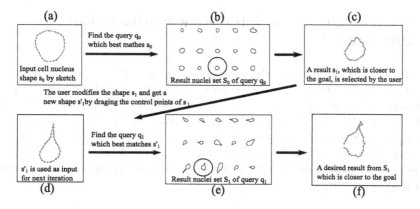

Fig. 5. An example of nuclei shape query to illustrate the process of our method.

- **Step 2:** Algorithms are used to automatically generate the SQL query q_0 that matches the user's input best. Using q_0, database can return a set of shapes S_0 (Fig. 5(b)) matching the query. The query results from the dataset are returned to the user. The results may not fit the user's target perfectly, however, they can be regarded as feedback to improve the query in the next iteration.
- **Step 3:** The user can select a candidate result $s_1 \in S_0$ (Fig. 5(c)) that best matches his target. Since it may be different from the user's expectation. Some modifications to s_1 can be made through visual interface. In this example, the shape is adjusted by dragging the control points of the nucleus polygons. In this way more features can also be added to s_1. The new shape after modification is s'_1 (Fig. 5(d)). Then the user can use s'_1 as input again for a new result set S_1 (Fig. 5(e)), which will be a subset of S_0.

The user repeats from Step 1 to Step 3 until a satisfied nucleus shape is found.

4.3   Locality Sensitive Hashing

The pathology image may contain million or hundreds of millions of cells. When searching a specific shape, we need to scan the all the cells and compute the similarity with the given shape. The time complexity of this process is $O(n)$, which is extremely computationally expensive and hard to be finished in real time. In order to solve this issue, we apply locality sensitive hashing (LSH) technique to lower down the time complexity near $O(logn)$. Specifically, it reduces the dimensionality of high-dimensional data and hashes input items so that similar items map to the same buckets with high probability (the number of buckets being much smaller than the input items). LSH has much in common with data clustering and nearest neighbor search. It approximates the nearest k neighbors through solving a series of sensitive tasks. The overall time complexity is near $O(logn)$, when k is not large.

5   Window Query for Interesting Nuclei Density Distributions

The density change of nuclei distribution is critical for the analysis of cancer region. Although finding interesting regions can be finished automatically by computer vision algorithms, a medical expert still needs to verify the regions in details to reduce the uncertainty of the algorithm results. Furthermore, when existing knowledge is unclear about the density distribution of the cancer region, the medical expert will have to explore the images for decision making. In this section, we propose a method for exploring and searching density distribution with a specific pattern.

5.1   Constructing Density Field of Nuclei

Since each image in the pathology dataset contains extremely large dimensions, such as $200\,k \times 200\,k$, it is hard to deal with a density filed with such dimension. So we construct a smaller 2D matrix which represents the nuclei density distribution for all the image. The original images are divided into small tiles, with small resolution, such as 100×100. The number of nuclei in each tile is counted and stored in a 2D matrix M_1 of tiles. For this example, the matrix will be 2000×2000, with each element in the matrix represents a tile in the original image. The matrix can be viewed as a 2D scalar field of the nuclei density distribution. By this way, we can reduce the complexity of the density matching to find a window with the desired distribution.

5.2 Input Density Distribution by Sketching

The user can generate their desired 2D distribution by sketching lines on the canvas (Fig. 6(a)). For each point in user's sketch, it can be regarded as a sample from the desired distribution f. Kernel density estimation is used to construct the distribution f (Eq. 1) of the user's input, where Gaussian function is employed as kernel distribution $K(x)$.

$$f(x) = \frac{1}{n} \sum_{p \in sketch}^{n} K_h(x - p) \tag{1}$$

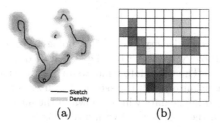

(a)   (b)

Fig. 6. (a) An example of density distribution generated by user sketch using kernel density estimation. (b) The discrete 2D density matrix generated from (a) for querying the windows with similar density distribution in images.

Here p is a point in the input sketch, n is the number of points in the sketch. Then the density distribution f is discretized into a 10×10 matrix m_1 (Fig. 6(b)). The values in the matrix m_1 represent the density in the 2D distribution. This matrix is used for matching similar distributions in the image. The user is able to set the dimension of the matrix m_2 to search distribution patterns in different scales.

5.3 Density Similarity Matching

Given a 10×10 matrix m with density values as input (Fig. 7(a)), a window with the same size is used to scan the matrix of tiles M_1. The distance between the window and input sketch is computed using Euclidean distance of matrix. The windows with distances smaller than a given threshold are ranked and returned (Fig. 7(c)). Matching the window with rotation will be considered in the future work.

The user can select an interesting window from the result to view the detailed image (Fig. 7(d)). The position of the window is indicated in the image overview with red box (Fig. 7(a)).

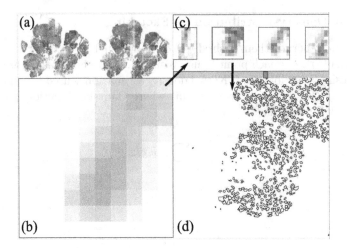

Fig. 7. The user interface of density query of pathology dataset. (a) The overview of an image from the pathology dataset. (b) The sketch of desired density distribution is drawn by the user. (c) The candidate windows of distributions in the image which have similar density distribution with the sketch. The user can select one of the result window, and the detailed image data will be shown in the nuclei view (d). The selected window is also shown in the overview (a) with a red box to indicate its position in the image. (Color figure online)

6   Window Query of Nuclei Growth Direction

Growth directions of cells are critical for detecting tumor region as well. For example, Fig. 1 indicates a tumor region, where the major directions of nuclei around the empty region are mainly pointing outwards. However, few existing works are able to query the direction patterns of nuclei in different regions. We propose a method to find windows with interesting direction patterns of nuclei in this section.

6.1   Constructing Direction Field of Nuclei Directions

Similar to the scalar field of density distribution in Sect. 5.1, the nuclei growth direction can be regarded as a vector field. Generally speaking, if the shape of a nucleus is close to a circle, it means that the cell grows evenly in all directions. We mainly focus on the cells growing in one direction, which can be a feature of tumor region (Fig. 1). For each nucleus, the growth direction $\vec{d}$ is estimated by the long axis direction $\vec{d'}$ of the nuclei shape polygon. However, $\vec{d}$ can have the same direction of $\vec{d'}$ or the opposite direction $-\vec{d'}$. So a vector field of nuclei direction is hard to be built unless the sign of direction d of each nucleus is certain.

We deal with this problem by defining a scalar direction field of nuclei in an image. The slope t of the major axis of a nucleus is computed. Since the nearby nuclei can have the same directions, the slope value can be continuous in neighbor regions in the image. The direction field is constructed for the image of nuclei (Fig. 8(a)) by the following two steps.

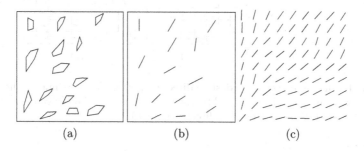

(a)   (b)   (c)

Fig. 8. (a) The shape polygons of nuclei in an image are extracted. (b) Find the major axis and its slope of each nucleus. (c) Calculate the direction field by interpolation using nearby nuclei directions.

Find the Major Axis of Each Nuclei. There are several options for deciding the major axis of a polygon. The minimum-area-rectangle can be found for each polygon. The long axis of the rectangle is regarded as the major axis. Another method for estimating the major axis is finding the longest chord inside the polygon of a nucleus. The longest chord can be found by enumerating all the pair of the vertices in each polygon. For the major axis of each nucleus, its slope value is calculated (Fig. 8(b)).

Estimate the Whole Direction Field. A matrix of tiles M_2 can be constructed using the same method as described in Sect. 5.1 for each image in the pathology dataset. Each element of the matrix (Fig. 8(c)) is x, and its position is $p(x)$. The values of it are calculated by all the nearby n nuclei $\{c_i\}$ within a radius r, i.e., $(p(x) - p(c_i))^2 < r^2$.

$$t(x) = \frac{1}{n} \sum_{c_i}^{n} t(n_i) K_h(p(x) - p(c_i)) \tag{2}$$

In Eq. 2, $t(x_i)$ is the slope of a nucleus c_i, K_h is the Gaussian function. The value of an element in the matrix can be regarded as the weighted average slopes of the nearby nuclei within a given radius r. As a result, the matrix M_2 represents the scalar field of the growth direction of the nuclei in the image.

6.2 Sketch-Based Query of Nuclei Directions

The user inputs the desired growth direction patterns by sketching the direction lines (Fig. 9). Each line on the input canvas represents the approximated direction of that part. A 10×10 matrix m_2 discretizing the canvas is constructed, which will represent the user's input pattern of directions on the canvas.

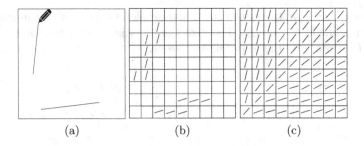

Fig. 9. (a) The user represents the desired direction pattern by sketching on the canvas. (b) A matrix of directions is initialized by the user's input. (c) All the value of elements in the matrix are estimated from the directions in (b).

For each element of m_2, if it intersects a line l drawn by the user, its value is initialized as the slope value $t(l)$ of that line l. Then all the element x of m_2 are estimated with Eq. 3 in a way similar to density field construction. In Eq. 3, $p(x) - p(y)$ measures the distance between the element x and a nearby element y in the matrix m_2.

$$t(x) = \frac{1}{n} \sum_{p \in m_2}^{n} t(p) K_h(p(x) - p(y)) \tag{3}$$

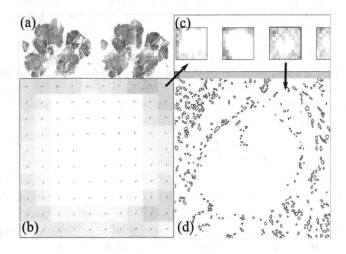

Fig. 10. The result of growth direction query of nuclei. (a) The overview of the image data. (b) The desired growth direction patterns of nuclei are generated from direction sketch by the user. The user also inputs the density sketch shown in the background. (c) The result view lists the windows which match both the density distribution and growth direction distribution. The user selects one of the windows, and the corresponding detailed nuclei data are shown in (d). The position of selected window is indicated as red box in (a). (Color figure online)

Similar to matching the density field in Sect. 5.3, a window of 10×10 is used to scan the matrix M_2. If a window has a small Euclidean distance with the input m_2, it will be put in the result set.

The density query and direction query can be merged together to find interesting patterns in both density and the directions of nuclei. It helps to filter out undesired regions. For example, the user generates a density query to find a region with empty inside (shown as the grey-scale heat map in Fig. 10(b)) as well as the direction query in which cells grow around the empty region (as shown in direction glyphs in Fig. 10(b)). The query result (Fig. 10(d)) shows a situation where the cells probably grow around a vessel.

7 Conclusion

We present an iterative visual query approach to search for interesting shapes, density distributions, and growth directions of nuclei in pathology databases. For future work, our approach needs to be tested with different cancer datasets for broader applications. We will involve pathologists to validate our methods. To make it more convenient for users, we will provide a tablet based interface for users to sketch queries, which will be executed on a backend server. Results from shape query can be filtered further with other nuclei properties such as skewness, area, and size. For density query, matching distribution patterns with rotation is needed for searching. More works can be done for building a vector field of growth directions in an image for direction query. In this way, regions in the vector field with high divergence can be calculated and listed as candidate cancer regions.

Acknowledgments. This research is supported in part by grants from National Science Foundation ACI 1443054, IIS 1350885 and IIS 1527200, National Institute of Health K25CA181503, the Emory University Research Committee, BNL LDRD grant 16-041, and the MSIP (Ministry of Science, ICT and Future Planning), Korea, under the "IT Consilience Creative Program (ITCCP)" (NIPA-2013-H0203-13-1001) supervised by NIPA.

References

1. Behrisch, M., Korkmaz, F., Shao, L., Schreck, T.: Feedback-driven interactive exploration of large multidimensional data supported by visual classifier. In: IEEE VAST, pp. 43–52. IEEE (2014)
2. Bimbo, A.D., Pala, P.: Visual image retrieval by elastic matching of user sketches. IEEE Trans. Pattern Anal. Mach. Intell. **19**(2), 121–132 (1997)
3. Caduff, D., Egenhofer, M.J.: Geo-Mobile queries: sketch-based queries in mobile GIS-Environments. In: Li, K.-J., Vangenot, C. (eds.) W2GIS 2005. LNCS, vol. 3833, pp. 143–154. Springer, Heidelberg (2005). doi:10.1007/11599289_13
4. De Silva, G.C., Yamasaki, T., Aizawa, K.: Sketch-based spatial queries for retrieving human locomotion patterns from continuously archived gps data. IEEE Trans. Multimedia **11**(7), 1240–1253 (2009)

5. Egenhofer, M.J.: Query processing in spatial-query-by-sketch. J. Vis. Lang. Comput. **8**(4), 403–424 (1997)
6. Eitz, M., Hildebrand, K., Boubekeur, T., Alexa, M.: Sketch-based image retrieval: benchmark and bag-of-features descriptors. IEEE Trans. Visual. Comput. Graph. **17**(11), 1624–1636 (2011)
7. Jiang, L., Mandel, M., Nandi, A.: Gesturequery: a multitouch database query interface. Proc. VLDB Endowment **6**(12), 1342–1345 (2013)
8. Latecki, L.J., Lakämper, R.: Shape similarity measure based on correspondence of visual parts. IEEE Trans. Pattern Anal. Mach. Intell. **22**(10), 1185–1190 (2000)
9. Parameswaran, A., Sarma, A.D., Garcia-Molina, H., Polyzotis, N., Widom, J.: Human-assisted graph search: it's okay to ask questions. Proc. VLDB Endowment **4**(5), 267–278 (2011)
10. Sarma, A.D., Parameswaran, A.G., Garcia-Molina, H., Widom, J.L.: Synthesizing view definitions from data. In: ICDT (2010)
11. Wang, F.: Pathology analytical imaging standards (PAIS) data management system. http://www.openpais.org/
12. Wang, F., Kong, J., Cooper, L., et al.: A data model and database for high-resolution pathology analytical image informatics. J. Pathol. Inform. **2**(1), 32 (2011)
13. Wang, F., Kong, J., Gao, J., et al.: A high-performance spatial database based approach for pathology imaging algorithm evaluation. J. Pathol. Inform. **4**(1), 5 (2013)
14. Wang, K., Huai, Y., Lee, R., Wang, F., Zhang, X., Saltz, J.H.: Accelerating pathology image data cross-comparison on CPU-GPU hybrid systems. Proc. VLDB Endowment **5**(11), 1543–1554 (2012)
15. Wattenberg, M.: Sketching a graph to query a time-series database. In: CHI 2001 Extended Abstracts on Human factors in Computing Systems, pp. 381–382. ACM (2001)
16. Wei, J., Wang, C., Yu, H., Ma, K.-L.: A sketch-based interface for classifying and visualizing vector fields. In: IEEE Pacific Visualization Symposium (PacificVis), pp. 129–136. IEEE (2010)
17. Yang, J., Hubball, D., Ward, M.O., Rundensteiner, E.A., Ribarsky, W.: Value and relation display: interactive visual exploration of large data sets with hundreds of dimensions. IEEE TVCG **13**(3), 494–507 (2007)
18. Zhu, J.-Y., Lee, Y.J., Efros, A.A.: AverageExplorer: interactive exploration and alignment of visual data collections. ACM Trans. Graph. (TOG) **33**(4), 160 (2014)

Storing and Querying DICOM Data
with HYTORMO

Danh Nguyen-Cong[1]([✉]), Laurent d'Orazio[1], Nga Tran[2],
and Mohand-Said Hacid[3]

[1] LIMOS Laboratory, UMR 6158 CNRS,
Blaise Pascal University Clermont-Ferrand II, 63173 Aubière, France
{nguyenda, dorazio}@isima.fr
[2] HPE Vertica, Cambridge, MA, USA
nga.tran@hpe.com
[3] LIRIS – University of Claude Bernard Lyon 1,
43, boulevard du 11 Novembre 1918, 69622 Villeurbanne, France
mshacid@liris.univ-lyonl.fr

Abstract. In the health care industry, DICOM (Digital Imaging and Communication in Medicine) standard has become very popular for storage and transmission of digital medical images and reports. The ever-increasing size, high velocity and variety of the DICOM data collections make them more and more inefficient to be stored and queried them using a single data storage technique, e.g., a row store or a column store. In this study, we first highlight challenges in DICOM data management. We then describe HYTORMO, a new model to store and query the DICOM data. HYTORMO uses a hybrid data storage strategy that is aimed not only to leverage the advantage of both row and column stores, but also to attempt to keep a trade-off among reducing disk I/O cost, reducing tuple construction cost and reducing storage space. In addition, Bloom filters are applied to reduce network I/O cost during query processing. We prototyped our model on the top of Spark. Our preliminary experiments validate the proposed model in real DICOM datasets and show the effectiveness of our method.

Keywords: DICOM · Medical image data · Hybrid store · Bloom filter

1 Introduction

In the health care industry, the management of ever-increasing volumes of medical image data becomes a real challenge. The development of imaging technologies, the long-term retention of medical data imposed by medical laws and the increase of image resolution are all causing a tremendous grow in data volume. In addition, the different acquisition systems (Philips, Olympus, etc.) to be used, the distinct specialties (gastroenterology, gynaecology, etc.) to be considered as well as preferences of physicians, nurses or other health-care professionals lead to a high variety, even if data follow the widely adopted DICOM (Digital Imaging and Communication in Medicine) standard [1]. The huge volume, high velocity and variety of the medical image data make this

© Springer International Publishing AG 2017
F. Wang et al. (Eds.): DMAH 2016, LNCS 10186, pp. 43–61, 2017.
DOI: 10.1007/978-3-319-57741-8_4

domain a concrete example of Big Data [2]. In this paper, we focus on storing and querying medical image data that follow the DICOM standard.

With the widely use of the DICOM standard nowadays, there have been some studies on DICOM data management [3–6]. However, complex characteristics of DICOM data make efficient storing and querying non-trivial tasks. The proposed solutions limited themselves to query types and could have negative impacts on performance and scalability. Most state-of-the-art DICOM data management systems employ traditional row-oriented databases ("row-RDBMS/row stores"). Using a row-RDBMS typically requires a query processor to read the entire database table into memory. This causes a lot of unnecessary disk I/O even when only a few attributes are used. In addition, the current studies have not introduced solutions to reduce a large amount of useless intermediate results created during query processing even if the final result is very small. As a result, in large-scale distributed database systems, running a computing cluster, these useless intermediate results can generate a lot of unnecessary network I/O to exchange data between cluster nodes.

In recent years, some studies have already proposed read-optimized databases to avoid reading unnecessary data from query processing. The read-optimized databases can include either column-oriented databases ("column-RDBMS/column stores"), such as MonetDB [7] and C-Store [8], or hybrid row/column-oriented databases, such as Fractured Mirrors [9], HYRISE [10], and SAP HANA [30]. The advantage of these databases is to reduce disk I/O cost. However, their tuple reconstruction cost is high and thus cannot cope with the high heterogeneity and evolution of DICOM data.

On the other side, cloud-based systems have provided solutions of high performance computing together with reliable and scalable storage to facilitate growth and innovation at lower operational costs. Hadoop [23] has become one of the de facto industry standards in this area. MapReduce has also shown a very good scalability for batch-oriented data processing. However, since they are designed for general-purpose, the major challenge is how to build a specialized system to effectively store and query DICOM data.

In this paper, we introduce a novel model, called HYTORMO, to store and query DICOM data. It provides a novel hybrid data storage strategy using both row and column stores to avoid reading unnecessary data as well as to reduce tuple construction cost and storage space. In addition, Bloom filters [11] are integrated into query processing to reduce intermediate results in join sequences.

Our major contributions can be summarized as follows: (1) We determine the characteristics of DICOM data that cause challenges in data management. (2) We propose a hybrid data storage strategy using both row and column stores to reduce disk I/O cost, tuple reconstruction cost and storage space. (3) We provide a query processing strategy with Bloom filters to reduce network I/O cost. (4) We finally present preliminary experiments with real DICOM data to show the effectiveness of our approaches.

The rest of this paper is organized as follows. Section 2 highlights problem definition. Section 3 describes the architecture of HYTORMO and the details of its components. Section 4 presents preliminary experimental results. Section 5 discusses related works. Finally, we conclude the paper and give an outlook on future works in Sect. 6.

2 Problem Definition

In this Section, we introduce DICOM standard, its challenges in data management, current database techniques, and problem formulation of our study.

2.1 DICOM Standard and Its Challenges

The DICOM standard was initially developed in 1983 by a joint committee of American College of Radiology and the National Electrical Manufacturers Association [1]. After many changes, in 1993 DICOM Version 3.0 was published to be widely used. The primary objective of this standard is to define data layouts and exchange protocols for storage and transmission of digital medical images and reports between medical imaging systems. Here we are mainly interested in data in DICOM files. The structure of a DICOM file is divided into three portions: (i) *a header (to recognize if it is a DICOM file)*, (ii) *metadata (to store information related to the image)*, and (iii) *pixel data (to store the actual image pixels)*. The metadata contains attributes which encode attributes of real-world entities (Patients, Studies, Series, etc.) related to the image. For instance, information about Patient is stored in attributes such as Name, Identity Number, Date of Birth, and Ethnic Group.

The following characteristics of DICOM data mainly cause challenges in data management: **Heterogeneous Schema.** The number of attributes in a DICOM file is very large, about 3000 attributes. Some of them are mandatory while others are optional. However, the number of attributes that are really used at a time varies dramatically depending on the availability of information acquired through performing a particular examination modality (CT, MRI, etc.) using a certain DICOM device (CT scanner, MRI scanner, etc.). **Evolutive Schema.** For instance, modalities or image acquisition devices are modified or added newly. **Variety.** Images and metadata. **Voluminous Data.** The storage space requirements of image databases are very large (e.g., terabyte) and ever-increasing tremendously. For instance, in France, information and tests results of a patient should be stored for up to 30 years [13].

So far, current solutions have provided limited supports to handle the above-mentioned characteristics. We present current database techniques in next Subsection.

2.2 Row- vs. Column-Oriented Databases

Most traditional databases (Oracle, SQL Server, etc.) are row-oriented databases that employ a row-oriented layout, moving horizontally across the table and storing attributes of each tuple consecutively on disk. This architecture is optimized for write-intensive online transaction processing (OLTP) because it is easily to add a new tuple and to read all attributes from a tuple. Their disadvantage is that if only a few attributes are accessed per query at once, the entire tuple still needs to be read into memory from disk before projecting. This wastes the I/O bandwidth [27]. Therefore row-oriented databases are not efficient in the case of highly heterogeneous data. In contrast, column-oriented databases (MonetDB, C-Store, etc.) are optimized for read-intensive workloads (OLAP). By storing data in columns rather than rows, only necessary attributes are read per a query. This saves I/O bandwidth [28]. However, their tuple reconstruction cost is higher than that cost of row-oriented databases.

2.3 MapReduce vs. Spark

Most of current row- and column-oriented databases have been developed to be used for structured data in relational database systems. They do not scale well and are ineffective to process semi/unstructured data. In contrast, MapReduce is originally developed to process extremely large amounts of semi/un/structured data. It provides a scalability and elasticity solution for Big Data. Unfortunately, batch processing data model of MapReduce is suitable for long running queries [29]. It does not support efficiently for users to execute interactive applications (e.g., ad hoc queries to explore data) because these applications have to share data (between parallel operations) across multiple steps of MapReduce and thus need overhead costs in both data replication and disk I/O. In contrast, Spark is an in-memory cluster computing system which can run on Hadoop [14]. Spark improves upon MapReduce by removing the need to write data to disk between steps. We are justified to use Spark due to its high performance for interactive queries and scalability.

2.4 Problem Formulation

Storing and querying DICOM data have been challenged by the complex characteristics of DICOM data (i.e., huge/ever-growing data size, variety, and heterogeneous/evolutive schema). We transform these problems into optimization problems of query performance and storage space. In this way, our study focuses on reducing I/O costs and storage space. We determine four main technical challenges: disk I/O cost, network I/O cost, tuple reconstruction cost, and storage space. We propose the HYTORMO model for efficient storing and querying DICOM data. The following design rules are considered to build this model:

- Reduce disk I/O, tuple reconstruction cost, and storage space by a hybrid data storage strategy using both row and column stores.
- Transparently rewrite user queries to access to data in row and column stores, without the need for any user intervention.
- Reduce network I/O cost by minimizing the intermediate results during query processing. An application of Bloom filters will help us to achieve this goal.

3 HYTORMO Architecture

HYTORMO is aimed to cope with heterogeneity, evolution, variety and huge volume of DICOM data. Its architecture consists of two components: *Centralized System* and *Distributed Nodes*, as shown in Fig. 1. The query processing is tightly integrated in

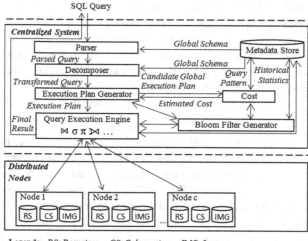

Fig. 1. The overall architecture of HYTORMO.

both Centralized System and Distributed Nodes. DICOM data are distributedly stored in nodes of Distributed Nodes (using both row and column stores).

3.1 Data Storage Strategy

The main goals of data storage strategy are to reduce disk I/O, tuple reconstruction cost, and storage space. *Metadata* and *pixel data* of DICOM files are extracted and stored in Hadoop distributed file system (HDFS) in a manner to achieve these goals. However, in the scope of this paper, we mainly concern about storing the metadata portion as the full-content images can be easily accessed from the metadata via links.

We propose a hybrid data storage strategy using both row and column stores for the metadata. Due to the complexity characteristics of DICOM data, identifying which attributes should be put in a particular store is a challenge work. We present a novel vertical data partitioning schema that is based on attribute classification.

First, we create entities and use an entity-relationship (ER) model to represent the logical relationships between the entities such as Patient, Study, Series, GeneralInfoTable, SequenceAttributes, and Image. Each entity is described by a set of attributes, e.g., *Patient(UID, Patient Name, Patient ID, Patient Birth Date, Patient Sex, Ethnic Group, Issuer Of Patient ID, Patient Birth Time, Patient Insurance Plan Code Sequence, Patient Primary Language Code Sequence, Patient Primary Language Modifier Code Sequence, Other Patient IDs, Other Patient Names, Patient Birth Name, Patient Telephone Numbers, Smoking Status, Pregnancy, Last Menstrual Date, Patient Religious Preference, Patient Comments, Patient Address, Patient Mother Birth Name, Insurance Plan Identification)*, where *UID* is an unique identifier. The ER model then is converted into a relational data model that consists of relations.

Second, attributes of each relation will be classified to fall into one of three categories: *(1) Mandatory:* Attributes are not allowed to get *null* values and are frequently-accessed-together. *(2) Frequently-accessed-together:* Attributes are allowed to get *null* values and frequently accessed together. *(3) Optional/Private/Seldom-accessed:* Attributes are allowed to get *null* values and not frequently accessed together (for short, we sometimes call them *"Optional"*).

Finally, attributes of the same category will be stored in the same table as below:

- Attributes of the first two categories are stored in tables of row stores, called "row tables". The aim is to reduce tuple reconstruction cost. For instance, *Patient Name, Patient ID, Patient Birth Date, Patient Sex,* and *Ethnic Group* are classified as "Mandatory" and stored in a row table. *Pregnancy* and *Last Menstrual Date* are classified as "Frequently-accessed-together" and stored in another row table.
- Attributes of the last category are stored in tables of column stores, called "column tables". The aim is to save the I/O bandwidth if only a few attributes are accessed per query at once. For instance, the rest of attributes of the Patient entity are classified as "Optional" and stored in a column table.

The above vertical data partitioning schema is non-overlapping, that is an attribute only belongs to a table except *UID*. In addition, to reduce storage space, we do not store rows with only *null* values.

3.2   Query Processing Strategy

The goal of query processing strategy can be briefly described as follows: It is given that DICOM data are distributedly stored across row and column stores. Find a query processing strategy to minimize the intermediate results.

Global Description of the Strategy. The actual query processing includes query parsing, query decomposition, query optimization, and query execution. These phases are shown in Fig. 1. The query is parsed by the Parser. It then is decomposed into sub-queries by Decomposer. The query decomposition increases the efficiency of the query by directing sub-queries only to the corresponding row and column tables that contain the required data, leading to a significant reduction of query input size. This also allows HYTORMO to utilize benefits of both row and column stores. The query optimization is performed by Execution Plan Generator that evaluates possible execution plans (i.e., different join strategies for combing results of sub-queries) and chooses the one with minimum cost. Since a given query could have a large number of execution plans due to different join ordering possibilities, an exhaustive search for an optimal execution plan is too expensive. We thus adopt to use a *left-deep sequential tree plan* introduced by Steinbrunn et al. [15]. In this plan, a join that yields a smaller intermediate result will be computed first. Metadata Store keeps metadata of database tables (schemas, cardinality of tables, etc.) that can be used during query processing. Finally, Query Execution Engine processes the query execution plan. It sends sub-queries to be executed on Distributed Nodes and retrieves intermediate results. In the end, it returns the final result to the front end. *Distributed Nodes* is mainly responsible for storing DICOM data and executing tasks that are assigned by the Centralized System. Bloom Filter Generator generates Bloom filters to remove irrelevant data out of inputs of joins if their benefits are found.

Query Decomposition. Our study focus on *Select-Project-Join (SPJ)* queries that involve selection conditions followed by equi-joins on surrogate attributes (UIDs) of row and column tables. In order to avoid loss of generality, we use a general form of SQL query to present a user query Q as given below.

Q: **SELECT** $T_I.UID^{RC}$, $T_I.att_a^{Rm}$, $T_I.att_b^{C}$, $T_J.att_x^{Rm}$, $T_J.att_y^{Rf}$, $T_K.att_z^{C}$
   FROM $\{T_I, T_J, T_K\}$
   WHERE $\{T_I.UID^{RC} = T_J.UID^{RC}\}$ **AND** $\{T_J.UID^{RC} = T_K.UID^{RC}\}$
   $\{T_I.att_a^{Rm} \theta\ value_a^{Rm}\}$ **AND** $\{T_I.att_b^{C} \theta\ value_b^{C}\}$ **AND**
   $\{T_J.att_x^{Rm} \theta\ value_x^{Rm}\}$ **AND** $\{T_K.att_z^{C} \theta\ value_z^{C}\}$

   where:

   o T_I, T_J, T_K: entity tables
   o $T_I(UID^{RC}, att\_{}^{Rm}, ..., att\_{}^{Rf}, ..., att\_{}^{C}, ...)$: schema of T_I
   o $T_J(UID^{RC}, att\_{}^{Rm}, ..., att\_{}^{Rf}, ..., att\_{}^{C}, ...)$: schema of T_J
   o $T_K(UID^{RC}, att\_{}^{Rm}, ..., att\_{}^{Rf}, ..., att\_{}^{C}, ...)$: schema of T_K
   o $att\_{}^{Rm}$: a mandatory attribute is stored in a row table
   o $att\_{}^{Rf}$: a frequently-accessed-together attribute is stored in a row table
   o $att\_{}^{C}$: an optional/private/seldom-accessed attribute is stored in a column table
   o $value\_{}^{Rm}$, $value\_{}^{Rf}$, $value\_{}^{C}$: constant values
   o θ: one of $\{<, \le, =, >, \ge,$ **LIKE, NOT LIKE**$\}$

We use superscripts *Rm, Rf,* and *C* to indicate that the corresponding attribute will be stored in a row table of mandatory attributes, a row table of frequently-accessed-together attributes, or a column table of optional/private/seldom-accessed attributes, respectively. A superscript *RC* is to indicate that the corresponding attribute is stored in both row and column tables. However, these superscripts are not shown to the user.

The plan tree of query Q is given in Fig. 2(a). Here, T_I, T_J, and T_K are entity tables whose names, *e.g., Patient, Study, Series, etc.,* are used in Q by the user. We assume that each of these tables, has been vertically partitioned into several "child" tables, i.e., row and column tables, by applying the data storage strategy presented in Sect. 3.1. However, only some of the child tables are required by Q. We also assume that Q is decomposed into sub-queries Q_I, Q_J, and Q_K that are further decomposed into smaller sub-queries $Q_{I,1}$, $Q_{I,2}$, $Q_{J,1}$, $Q_{J,2}$, and $Q_{K,1}$ to able to directly map to child tables containing required attributes. As presented in Fig. 2(b), $Q_{I,1}$ and $Q_{I,2}$ access to T_1 and T_2, respectively, that are child tables of T_I. Similarly, $Q_{J,1}$ and $Q_{J,2}$ access to T_3 and T_4, respectively, that are child tables of T_J. $Q_{K,1}$ only accesses to T_N, a child table of T_K.

HYTORMO uses a *left-deep sequential tree plan* for joining intermediate results of sub-queries. It will automatically determine a join as an inner or a left outer join. Because Q is a user query, entity tables are used in Q. Thus the type of a join between two entity tables is explicitly identified by the user. For instance, Q in Fig. 2(a) can be

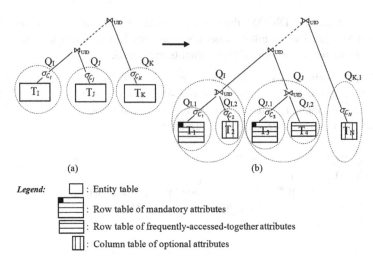

Fig. 2. Plan tree of the SQL query

written as $Q = Q_I \bowtie_{UID} Q_J \bowtie_{UID} Q_K$, only using inner joins. However, it is necessary to evaluate some joins between sub-queries as left outer joins to prevent data loss caused by the tuples discarded by inner joins.

We consider two cases in which a left outer join should be used. First, in a join between two child tables of the same entity table, if the left table is a row table of mandatory attributes while the right table is either a column table of optional attributes or a row table of frequently-accessed-together attributes, this join should be evaluated as a left outer join. For instance, in Fig. 2(b), both sub-queries $Q_{I,1} \bowtie_{UID} Q_{I,2}$ and $Q_{J,1} \bowtie_{UID} Q_{J,2}$ will evaluated as left outer joins. This is because $Q_{I,1}$ and $Q_{J,1}$ are mapped to row tables of mandatory attributes T_1 and T_3, respectively, while $Q_{I,2}$ and $Q_{J,2}$ are mapped to a column table of optional attributes T_2 and a row table of frequently-accessed-together attributes T_4, respectively. Second, in a join between two entity tables, if the right table has been changed to either a column table of optional attributes or a row table of frequently-accessed-together attributes, this join should be evaluated as a left outer join. For instance, in Fig. 2(b), $Q_{K,1}$ is mapped to column table of optional attributes, thus the join using the result of $Q_{K,1}$ is rewritten to a left outer join.

In the scope of this paper, we concern on inner joins and the two above-mentioned cases of left outer joins. To improve performance of a query, we need to reduce the number of left outer joins and to apply Bloom filters.

Reducing the Number of Left Outer Joins. We use the following heuristic rule for deciding whether or not a left outer join should be rewritten as an inner join: *Given a left outer join $T_1 \bowtie_{UID} T_2$, if the right table T_2 does not contain non-null constraints on its attributes, the left outer join is kept no change. In contrast, if the right table T_2 contains non-null constraints on its attributes, the left outer join should be rewritten as an inner join that might improve query performance.*

The above heuristic rule is based on the fact that, in the left outer join $T_1 \bowtie_{UID} T_2$, if T_2 does not contain non-null constraints on its attributes, the left outer join returns all matching tuples between T_1 and T_2, like an inner join. The unmatched tuples are also preserved from T_1 and are supplied with *nulls* from T_2. Thus, in this case, the left outer join is kept no change. However, if T_2 contains non-null constraints on its attributes, these constraints must be evaluate to TRUE to form a tuple in the result. They also eliminate any *nulls* of attributes from T_2. In this case, a left outer join is unnecessary, the left outer join thus should be rewritten as an inner join.

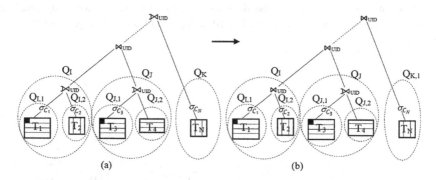

Fig. 3. Rebuilding the plan tree after reducing the number of left outer joins

Figure 3(a) presents a plan tree that has been introduced in Fig. 2. To apply the above heuristic rule to this plan tree, we look at right tables of left outer joins. Assume that σ_{C_2} and σ_{C_N} are non-null constraints on attributes of tables T_2 and T_N, respectively. Then left outer joins in $Q_{I,1} \bowtie_{UID} Q_{I,2}$ and $(Q_I \bowtie_{UID} Q_J) \bowtie_{UID} Q_{K,1}$ are rewritten as inner joins $Q_{I,1} \bowtie_{UID} Q_{I,2}$ and $(Q_I \bowtie_{UID} Q_J) \bowtie_{UID} Q_{K,1}$, respectively, as given in Fig. 3(b).

Application of Bloom Filters. Bloom filter (BF) is a space-efficient probabilistic data structure with little error allowable when used to test whether an element is a member of a set [11]. In our case, we consider to apply an intersection Bloom filter (IBF) rather than BFs because of its benefit in removing irrelevant data, as presented in [12]. The way how to apply an *IBF* to HYTORMO is given below.

We consider a general form of queries in HYTORMO. Assume that a query Q can be decomposed into a set of sub-queries $Q_1, Q_2, \ldots, Q_K$, each of which can be further decomposed into smaller sub-queries to able to map to input tables, i.e., row- and column-oriented tables $T_1, T_2, \ldots, T_N$. Q is in form of a multi-way join on common join attributes. Because HYTORMO uses a *left-deep sequential tree plan*, we focus on the application of the *IBF* for this plan.

Although input tables $T_1, T_2, \ldots, T_N$ might have some common join attributes, in the scope of this paper, we assume these tables only share the common join attribute *UID*. In this case, we can build a common *IBF* on the join attribute *UID* of a subset of the input tables. After built, the *IBF* can be probed to filter these input tables.

The build and probe phases of the *IBF* are illustrated in Fig. 4(a) and (b), respectively. We assume that the heuristic rule has been applied to reduce the number of left outer joins to obtain a plan tree (see Fig. 4(a)). In the build phase, we first compute a set of *BFs* of the same size and the same hash functions on the join attribute *UID* for intermediate result tables $D_1, D_2, ..., D_N$ that have been created as results of sub-queries $Q_{I,1}, Q_{I,2}, Q_{J,1}, Q_{J,2}$, and $Q_{K,I}$. We use DataFrames [14] of Spark to store these intermediate result tables. The *IBF* then is computed by bitwise *ANDing* the *BFs*. It is worthy to note that we do not compute a *BF* for the right table of a left outer join if it does not contain non-null constraints on its attributes. For instance, we do not compute a *BF* for D_4 (intermediate result table of $Q_{J,2}$) because there do not exist non-null constraints in $Q_{J,2}$ (i.e., $Q_{J,1} \bowtie_{UID} Q_{J,2}$ is not equivalent to $Q_{J,1} \bowtie_{UID} Q_{J,2}$). Thus, building a *BF* for D_4 can cause data loss caused by tuples discarded by *ANDing* this *BF* with others. The *IBF* is probed to filter input tables before a join occurs (see Fig. 4(b)).

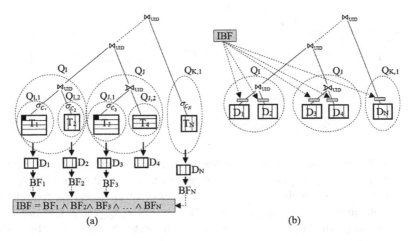

Fig. 4. Build (a) and Probe (b) phases of an IBF.

Cost Effectiveness of IBF. Our goal is to evaluate the effect of using or not using an *IBF* to query performance.

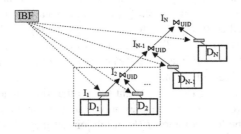

Fig. 5. The approximate left-deep sequential tree plan with the *IBF*.

For the sake of simplicity, we present an approximate cost estimate for the *IBF* on a sequential join sequence of N tables. For this approximation, we assume *BFs* are computed for all input tables $D_1, D_2, ..., D_N$. The *IBF* is computed from these *BFs* and probed to filter all of these input tables. We also assume that the sequential join sequence only includes inner joins, as illustrated in Fig. 5. Let that the multi-way join operation be $Q = D_1 \bowtie_{UID} D_2 \bowtie_{UID} ... \bowtie_{UID} D_N$, where $|D_i| \le |D_{i+1}|$, for every $i \in [1, N-1]$. The sequential join sequence for the left-deep sequential tree plan is:

$$Q = ((((D_1 \bowtie_{UID} D_2) \bowtie_{UID} ...) \bowtie_{UID} D_{N-1}) \bowtie_{UID} D_N).$$

Tables $D_1, D_2, ..., D_N$ and intermediate result tables $I_1, I_2, ..., I_{N-1}$ are used as inputs of joins. Here, we are setting $I_1 = D_1$ and $I_N = final\ query\ result$.

The performance of a multi-way join in a cluster is usually determined by *network I/O cost* and *disk I/O cost*. We thus use these costs to analysis the effectiveness of the *IBF*. We start this work by giving definitions and basic mathematical concepts. Assume that *BFs* and *IBF* have been built from tables $D_1, D_2, ..., D_{N-1}$ with the same configuration: using a vector of m bits and k hash functions $h_1(v), h_2(v), ..., h_k(v)$, where v is a value of join attribute. Table 1 shows notations used in cost models.

Table 1. Notations of cost models.

Notation	Description
D_i	Table is used as either a build or a probe table
I_i	Intermediate result table of sequential join sequence
BF_i	Bloom filter is built on table D_i
IBF	Intersection Bloom filter
ρ_{D_i}	Selectivity of table D_i
ρ_{BF_i}	Selectivity of Bloom filter BF_i that is associated to table D_i
ρ_{IBF}	Selectivity of IBF that is built on tables $D_1, D_2, ..., D_N$
P_{BF_i}	False positive of BF_i of table D_i due to hash collisions
P_{IBF}	False positive of IBF that is built on tables $D_1, D_2, ..., D_N$

The *probability of a false positive* of a Bloom filter BF_i due to hash collisions is calculate by (1) [16], where BF_i is representing a set of n_i values of the join attribute *UID* of table D_i in a vector of m bits and using k independent hash functions.

$$P_{BF_i} = \left(1 - \left(1 - m^{-1}\right)^{kn_i}\right)^k \approx \left(1 - e^{kn_i/m}\right)^k. \tag{1}$$

We define the *selectivity of Bloom filter* BF_i of table D_i in (2).

$$\rho_{BF_i} = \rho_{D_i} + \left(1 - \rho_{D_i}\right).P_{BF_i} \tag{2}$$

where $\left(1 - \rho_{D_i}\right).P_{BF_i}$ is the fraction of tuples from the probe table D_i that are not discarded by BF_i and do not join with any tuples in the build table.

Given the selectivity of Bloom filters of tables $D_1, D_2, ..., D_N$ that have been calculated by (2), the *selectivity of the IBF* is determined by (3).

$$\rho_{IBF} = \prod_{i=1}^{N} \rho_{BF_i}. \tag{3}$$

The *false positive of the IBF* can be calculated by (4).

$$P_{IBF} = \prod_{i=1}^{N} P_{BF_i} = \prod_{i=1}^{N} \left(1 - (1 - m^{-1})^{kn_i}\right)^{k}. \tag{4}$$

where N is the number of *BFs* with assumption that there exists a *BF* for each table D_i.

A comparison between (1) and (4) shows that value of P_{IBF} is less than value of P_{BF_i}. This means that applying an *IBF* will give a lower amount of false positive errors than only applying a single *BF*. The larger value of N, the smaller value of P_{IBF}.

In order to estimate *network I/O cost* and *disk I/O cost*, we depend on build and probe phases of the *IBF*, as given in Fig. 4(a) and (b), that include the following steps:

1. *Execute sub-queries to create intermediate result tables $D_1, D_2, ..., D_N$.*
2. *Compute $BF_1, BF_2, ..., BF_N$ on values of UIDs of $D_1, D_2, ..., D_N$, respectively.*
3. *Compute the IBF = $BF_1 \wedge BF_2 \wedge ... \wedge BF_N$.*
4. *Broadcast the IBF to all slave nodes of the cluster.*
5. *Apply the IBF to input tables $D_1, D_2, ..., D_N$ to obtain results $D_{1(filtered)}, ..., D_{N(filtered)}$.*
6. *Execute the sequential join sequence using tables $D_{1(filtered)}, ..., D_{N(filtered)}$ as inputs.*

The first three steps are in the build phase while the rest are in the probe phase. Assume that the first step has been done. We start to estimate costs from step 2.

Network I/O Cost. Since each join operation in the sequential join sequence will join an intermediate result table (created by the previous join) with an input table D_i. The *network I/O cost* when the *IBF* is not used, C_{Net}^{NoIBF}, can be calculated by (5).

$$C_{Net}^{NoIBF} = \sum_{i=1}^{N} size(D_i) + \sum_{i=1}^{N-1} size(I_i) \times size(D_{i+1}) \times \rho_{D_{i+1}, I_i}. \tag{5}$$

where $size(D_i)$ and $size(I_i)$ are size of input table D_i and intermediate result table I_i, respectively. ρ_{D_{i+1}, I_i} is selectivity factor (ratio of the joined tuples of D_{i+1} with I_i).

The cost C_{Net}^{NoIBF} consists of cost of sending input tables and intermediate result tables over the network. We assume that no replication is done on input tables.

The *network I/O cost* when the *IBF* is used, C_{Net}^{IBF}, can be computed by (6).

$$C_{Net}^{IBF} = c * size(IBF) + \sum_{i=1}^{N} size\left(D_{i(filtered)}\right)$$
$$+ \sum_{i=1}^{N-1} size(I_i) \times size\left(D_{i+1(filtered)}\right) \times \rho_{D_{i+1}(filtered), I_i}. \tag{6}$$

where c is the number of slave nodes of the cluster.

The cost C_{Net}^{IBF} consists of cost of sending (broadcast) the *IBF* to all of slave nodes of the cluster and cost of sending filtered input tables and intermediate result tables over the network. Here, we do not apply the *IBF* to filter intermediate results.

A comparison between (5) and (6) shows that $c * size(IBF)$ is usually small and $size\left(D_{i(filtered)}\right) \ll size(D_i)$. Therefore C_{Net}^{IBF} is less than C_{Net}^{NoIBF}.

Disk I/O Cost. The *disk I/O cost* without using *IBF*, $C_{I/O}^{NoIBF}$, can be calculated by (7).

$$C_{I/O}^{NoIBF} = \sum_{i=1}^{N-1} \left[size(I_i) + size(D_{i+1}) \right] + \sum_{i=2}^{N} size(I_i). \tag{7}$$

where:

- $\displaystyle\sum_{i=1}^{N-1} \left[size(I_i) + size(D_{i+1}) \right]$: reading intermediate results and input tables for joins.
- $\displaystyle\sum_{i=2}^{N} size(I_i)$: writing intermediate results to disk (here we are setting $I_1 = D_1$).

When the *IBF* is used, the *disk I/O cost*, $C_{I/O}^{IBF}$, can be calculated by (8).

$$\begin{aligned}
C_{I/O}^{IBF} = {}& 2 \times \sum_{i=1}^{N} size(D_i) + \sum_{i=1}^{N} size\left(D_{i(filtered)}\right) \\
& + \sum_{i=1}^{N-1} \left[size(I_i) + size\left(D_{i+1(filtered)}\right) \right] + \sum_{i=2}^{N} size(I_i).
\end{aligned} \tag{8}$$

where:

- $2 \times \sum_{i=1}^{N} size(D_i)$: reading the input tables two times (to build and to apply the *IBF*).
- $\sum_{i=1}^{N} size\left(D_{i(filtered)}\right)$: writing filtered input tables to disk after applying the *IBF*.
- $\sum_{i=1}^{N-1} \left[size(I_i) + size\left(D_{i+1(filtered)}\right) \right]$: reading intermediate results and filtered input tables to be used as inputs of joins.
- $\sum_{i=2}^{N} size(I_i)$: writing intermediate results to disk (here we are setting $I_1 = D_1$).

We assume that the *BFs* and the *IBF* are small enough to be stored in internal memories of slave nodes so that no disk I/O cost is needed for them. A comparison between (7) and (8) shows that $C_{I/O}^{IBF}$ includes extra costs to read and to write input tables during build and probe phases. However, then join operations will use filtered tables as their inputs. Therefore, if $size\left(D_{i(filtered)}\right) \approx size(D_i)$, there is no benefit when applying the *IBF*. However, if $size\left(D_{i(filtered)}\right) \ll size(D_i)$, we can achieve $C_{I/O}^{IBF} \approx C_{I/O}^{NoIBF}$.

4   Preliminary Experimental Results

This Section presents preliminary experimental results. We first describe experimental environment, datasets and experimental query. We then compare query performance against various storage strategies and effectiveness of *IBF*.

4.1   Experimental Environment

We have used Hadoop 2.7.1, Hive 1.2.1 and Spark 1.6.0 to create a cluster of seven different nodes (one master node and six slave nodes). The hardware of each node is the same and has the following configuration: Intel(R) core(TM) i7-3770 CPU @ 3.40 GHz, 16 GB RAM and 500 GB hard disk. We use the standard configuration with a modification: we change the replication factor of HDFS from 3 to 2 in order to save space. We implement the execution plan for the experimental query using Spark [14].

4.2   Datasets

We have used a mixed DICOM dataset of [17–21]. The metadata and pixel data are extracted from DICOM files using the library dcm4che-2.0.29 [22]. The attributes of metadata are classified and stored in a fashion as discussed in Sect. 3.1. We use sequence files and ORC files in Hive [23] to store row and column tables, respectively. A statistic of the DICOM datasets are given in Table 2.

Table 2. DICOM datasets used in the experiment.

Datasets	Number of files	Number of extracted attributes	Size of extracted metadata	Total size of files
CTColonography [17]	98,737	86	7.76 GB	48.6 GB
Dclunie [18]	541	86	86.0 MB	45.7 GB
Idoimaging [19]	1,111	86	53.9 MB	369 MB
LungCancer [20]	174,316	86	1.17 GB	76.0 GB
MIDAS [21]	2,454	86	63.4 MB	620 MB

Although there are many entities for DICOM data, below we only present schemas of entities required in the experimental query. The superscripts *Rm*, *Rf*, *C* and *RC* are the same as mentioned in Sect. 3.2.

– **Patient**(UID^{RC}, $PatientName^{Rm}$, $PatientID^{Rm}$, $PatientBirthDate^{Rm}$, $PatientSex^{Rm}$, $EthnicGroup^{Rm}$, $IssuerOfPatientID^{C}$, $PatientBirthTime^{C}$, $PatientInsurancePlanCode Sequence^{C}$, $PatientPrimaryLanguageCodeSequence^{C}$, PatientPrimaryLanguage $ModifierCodeSequence^{C}$, $OtherPatientIDs^{C}$, $OtherPatientNames^{C}$, PatientBirth $Name^{C}$, $PatientTelephoneNumbers^{C}$, $SmokingStatus^{C}$, $Pregnancy^{Rf}$, LastMenstrual $Date^{Rf}$, $PatientReligiousPreference^{C}$, $PatientComments^{C}$, $PatientAddress^{C}$, Patient-$MotherBirthName^{C}$, $InsurancePlanIdentification^{C}$)

Table 3. Row and column tables are used by hybrid data storage strategy.

Entity	Row table of "Rm" attributes	Row table of "Rf" attributes	Column table of "C" attributes
Patient	RowPatient	RowPregnancy	ColPatient
Study	RowStudy	–	ColStudy
GeneralInfoTable	–	–	ColGeneralInfoTable
SequenceAttributes	RowSequenceAttributes	–	–

- **Study**(UID^{RC}, StudyInstanceUID^{Rm}, StudyDate^{Rm}, StudyTime^{Rm}, ReferringPhysicianName^{Rm}, StudyID^{Rm}, AccessionNumber^{Rm}, StudyDescription^{Rm}, PatientAge^{C}, PatientWeight^{C}, PatientSize^{C}, Occupation^{C}, AdditionalPatientHistory^{C}, MedicalRecordLocator^{C}, MedicalAlerts^{C})
- **GeneralInfoTable**(UID^{RC}, GeneralTags^{C}, GeneralVRs^{C}, GeneralNames^{C}, GeneralValues^{C})
- **SequenceAttributes**(UID^{RC}, SequenceTags^{Rm}, SequenceVRs^{Rm}, SequenceNames^{Rm}, SequenceValues^{Rm})

Table 3 shows the corresponding row and column tables that are used to store the above schemas. Because our experiment will compare the query performance against various storage strategies, these schemas also need to be stored in only row tables and only column tables.

4.3 Experimental Query

We measure the execution time of the query using three different storage strategies: row store only, column store only, and hybrid store (hybrid data storage strategy). We also validate the effect of using or not using the *IBF* to query performance.

The experimental query is given in Fig. 6(a). It is to retrieve the information stored in *X-ray* DICOM files of *men* who are *non-smoking*, greater than or equal to *x years old*. The query is based on TPC-H query 3 and 4 [24], but here we only focus on *SPJ* queries. The attributes used in *SELECT* and *WHERE* clauses are also marked by superscripts to indicate that they are being stored in row or a column tables. Five tables *RowPatient, ColPatient, ColStudy, ColGeneralInfoTable*, and *RowSequenceAttribute* are required by the query. The query is decomposed into four subqueries *sQ1, sQ2, sQ3*, and sQ4, as given in Fig. 6(b). The query processing strategy presented in Sect. 3.2 is applied to build a left-deep processing tree step by step while trying to keep intermediate results as small as possible.

4.4 Preliminary Query Performance

We ran the query for (i) *storing all tables in row stores (RS)*, (ii) *storing all tables in column stores (CS)*, and (iii) *storing all tables in the proposed hybrid store (HS)*.

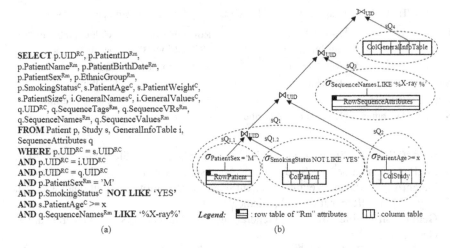

SELECT p.UID^RC, p.PatientID^Rm,
p.PatientName^Rm, p.PatientBirthDate^Rm,
p.PatientSex^Rm, p.EthnicGroup^Rm,
p.SmokingStatus^C, s.PatientAge^C, s.PatientWeight^C,
s.PatientSize^C, i.GeneralNames^C, i.GeneralValues^C,
q.UID^RC, q.SequenceTags^Rm, q.SequenceVRs^Rm,
q.SequenceNames^Rm, q.SequenceValues^Rm
FROM Patient p, Study s, GeneralInfoTable i,
SequenceAttributes q
WHERE p.UID^RC = s.UID^RC
AND p.UID^RC = i.UID^RC
AND p.UID^RC = q.UID^RC
AND p.PatientSex^Rm = 'M'
AND p.SmokingStatus^C **NOT LIKE** 'YES'
AND s.PatientAge^C >= x
AND q.SequenceNames^Rm **LIKE** '%X-ray%'

(a)

(b)

Fig. 6. The experimental query (a) and its execution plan tree (b)

The selectivity of the query vary depending on the predicates on attributes *PatientAge, PatientSex, SmokingStatus,* and *SequenceNames.* In our experiment, we vary the selectivity of the predicate on *PatientAge* to be *0.06 (PatientAge >= 90)*, *0.42 (PatientAge >= 60)*, and *0.88 (PatientAge >= 10)* but fix the selectivity of the others.

The chart in Fig. 7 shows that, for all cases of the selectivity, storing all schemas in only row stores or only column stores leads to higher execution time than that in a hybrid store. The rationale behind the query performance is in the own benefit of each data storage strategy. With the use of the hybrid store, storing mandatory attributes in row tables, e.g., *RowPatient* and *RowSequenceAttribute,* helped to reduce tuple reconstruct cost because most of these attributes are accessed together by the query. In contrast, only a few of optional/private/seldom-used attributes are required by the query. They thus should be stored in column tables, e.g., *ColPatient, ColStudy,* and *ColGeneralInfoTable* to save I/O bandwidth because only relevant attributes need to be read. If we store these attributes in row tables, the entire rows still have to be read from disk no matter how many attributes are accessed per query at once. This causes a waste of I/O bandwidth. Therefore, depending on a good understanding about the workload of regular queries, we can choose a right store for each attribute extracted from DICOM files to improve the query performance.

To evaluate the effect of using an *IBF,* we build an *IBF* from *BFs* that are computed for all tables except *ColGeneralInfoTable* since it is the right table of a left outer join. The accuracy of a *BF* is decided by ratio *m/n* where *m* is length of bit vector and *n* is size of set (i.e., cardinality of *UID* list of an input table). *m = 8n* has been considered a good balance between accuracy and space usage [24]. We thus apply this setting with *n* is the biggest cardinality value among tables (*RowPatient* in our case).

The chart in Fig. 8 gives a comparison between HYTORMO with (HS + IBF) and without (HS) using the *IBF.* In the best case, where the query is very highly selective (*PatientAge >= 90*), the IBF helped to reduce 80% of execution time, whereas in the worst case, where the query is lowly selective (*PatientAge >= 10*), the *IBF* increased

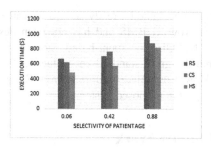

Fig. 7. Comparison of different storage strategies

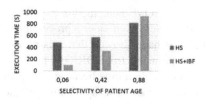

Fig. 8. A comparison of the effect of the IBF.

14% of the execution time. When *PatientAge* >= *60*, the *IBF* helped to reduce *41%* of the execution time. This is because in the best case the *IBF* removed a large amount of irrelevant tuples from joins. However, in the worst case, most of tuples of input tables are required in the final result and thus there are not much useless data to be removed by the *IBF*. The overhead costs incurred by build and probe phases of the *IBF* decrease query performance.

5   Related Works

There already exist several solutions to implement a DICOM data storage system. PACSs [25] mostly use row-RDBMSs to store, retrieve, and distribute medical image data. These systems are expensive but only support queries with predefined attributes and thus do not cope with heterogeneous schemas. eDiaMoND [2] stores DICOM data using a Grid-enabled medical image database that is built from row-RDBMSs. The system aims to provide inter-operability, scalability and flexibility. However, the development of query optimization techniques have not been introduced. Some commercial row-RDBMSs [4] have provided features to store and manage large-scale repositories of DICOM files. They add a new data type that enables any column of this type to store a DICOM content in their database table. Since a new separate object is created for each DICOM file, the storage space is quickly increased and thus decreases the overall performance of system. DCMDSM [5] is based on the original DSM [26] to vertically partition DICOM metadata into multiple small tables. The method is able to cope with the evolutive/heterogeneous schemas of DICOM data and saves I/O bandwidth. Unfortunately, the method already uses a centralized database approach developed on the top of a row-RDBMS and has not been designed to operate in a clustering environment. NoSQL document-based storage system [6] shares the schema-free non-relational design of standard key-value stores. It thus can handle the evolution of metadata. However, unlike traditional row-RDBMSs, there is no standardized query language for the proposed system.

6 Conclusion and Future Work

The high-performance DICOM data management becomes a real challenge. The current solutions still exist limitations to cope with the high heterogeneity, evolution, variety, and high volume of DICOM data. In this paper, we propose HYTORMO, using a hybrid data storage strategy and query processing strategy with *BFs*. Our preliminary experimental results have showed that it is necessary to carefully choose the right stores for attributes extracted from DICOM files. The use of both row and column stores results in lower execution time because it helps to reduce disk I/O, tuple reconstruction cost, and storage space. The application of the *IBF* helped to reduce network I/O cost because it removed irrelevant tuples out of inputs of joins. Our query performance is promising.

The next steps of our work is to reduce the overhead cost of *BFs*, a new cost model needs to be built to specify a threshold for selectivity factor of input tables so that *BFs* are only computed for tables that can be reduced large enough. Our future work also will include a comparison of HYTORMO to other methods such as commercial row-RDBMS [4] that use tables of a row-RDBMS to store schemas of metadata and use a single column of Object type in a table to store image content. SDSS SkyServer [31] also proposed a similar method, but to manage astronomy data. Furthermore, we consider to generate a bushy execution plan, instead of a left-deep tree plan.

References

1. Pianykh, O.S.: Digital Imaging and Communications in Medicine (DICOM): A Practical Introduction and Survival Guide. Springer, Heidelberg (2008)
2. Merelli, I., et al.: Managing, analysing, and integrating big data in medical bioinformatics: open problems and future perspectives. BioMed. Res. Int. 1–13 (2014)
3. Power, D., Politou, E., Slaymaker, M., Harris, S., et al.: A relational approach to the capture of DICOM files for grid-enabled medical imaging databases. In: SAC, pp. 272–279 (2004)
4. Annamalai, M., Guo, D., Susan, M., Steiner, J.: An oracle white paper: oracle database 11 g DICOM medical image support (2009)
5. Savaris, A., Härder, T., von Wangenheim, A.: DCMDSM: a DICOM decomposed storage model. J. Am. Med. Inform. Assoc. **21**, 917–924 (2014)
6. Rascovsky, S.J., et al.: Informatics in radiology: use of CouchDB for document-based storage of DICOM objects. Radiographics **32**, 913–927 (2012)
7. Boncz, P., et al.: MonetDB/X100: hyper-pipelining query execution. In: CIDR (2005)
8. Stonebraker, M., et al.: C-store: a column-oriented DBMS. In: VLDB, pp. 553–564 (2005)

9. Ramamurthy, R., DeWitt, D.: A case for fractured mirrors. VLDB **12**, 89–101 (2003)
10. Grund, M., et al.: HYRISE: a main memory hybrid storage engine. VLDB **4**, 105–116 (2010)
11. Bloom, B.H.: Space/time trade-offs in hash coding with allowable errors. Commun. ACM **13**, 422–426 (1970)
12. Phan, T.C., Orazio, L.D., Rigaux, P.: Toward intersection filter-based optimization for joins in MapReduce. In: Workshop Proceedings of the Cloud-I (2013)
13. OECD: Genetic Testing: A Survey of Quality Assurance and Proficiency Standards. OECD Publishing, Paris (2007)
14. Armbrust, M., et al.: Spark SQL: relational data processing in spark. In: SIGMOD (2015)
15. Steinbrunn, M., Moerkotte, G., Kemper, A.: Heuristic and randomized optimization for the join ordering problem. VLDB J. **6**, 191–208 (1997)
16. Broder, A., Mitzenmacher, M.: Network applications of bloom filters: a survey. Internet Math. **1**(4), 485–509 (2004)
17. CT Colonography. https://idash.ucsd.edu. Accessed 11 Oct 2015
18. David Clunie's Medical Image Format Site. http://www.dclunie.com. Accessed Oct 2015
19. Sample Data. http://idoimaging.com/wiki/. Accessed 12 Oct 2015
20. Lung Cancer Datasets. http://giveascan.org. Accessed 11 Oct 2015
21. MIDAS Datasets. http://www.insight-journal.org. Accessed 12 Oct 2015
22. Open Source Clinical Image and Object Management. http://www.dcm4che.org
23. White, T.: Hadoop: The Definitive Guide. 4th edn. O'Reilly Media, Inc., California (2015)
24. TPC-H specification 2.8.0. http://www.tpc.org/tpch/
25. Möller, M., Mukherjee, S.: Context-driven ontological annotations in DICOM images: towards semantic PACS. In: Proceedings of HEALTHINF (2009)
26. Copeland, G., Khoshafian, S.: A decomposed storage model. In: SIGMOD (1985)
27. Harizopoulos, S., et al.: Performance tradeoffs in read-optimized databases. In: VLDB (2006)
28. Floratou, A., Minhas, U.F., Özcan, F.: SQL-on-Hadoop: full circle back to shared-nothing database architectures. VLDB **7**, 1295–1306 (2014)
29. Popescu, A.D., Dash, D., Kantere, V., Ailamaki, A.: Adaptive query execution for data management in the cloud. In: CloudDB, pp. 17–24 (2010)
30. Rösch, P., Dannecker, L., Färber, F., Hackenbroich, G.: A storage advisor for hybrid-store databases. Proc. VLDB **5**(12), 1748–1758 (2012)
31. Szalay, A.S., et al.: The SDSS Skyserver: public access to the sloan digital sky server data. In: Proceedings of SIGMOD, pp. 570–581. ACM (2002)

Cloud-Based Whole Slide Image Analysis Using MapReduce

Hoang Vo[1], Jun Kong[2], Dejun Teng[3], Yanhui Liang[4], Ablimit Aji[5],
George Teodoro[6], and Fusheng Wang[4(✉)]

[1] Department of Computer Science,
Stony Brook University, Stony Brook, NY 11794, USA
hoang.v.vo@stonybrook.edu
[2] Department of Biomedical Informatics, Emory University, Atlanta, GA 30030, USA
[3] Department of Computer Science and Engineering,
The Ohio State University, Columbus, OH 43210, USA
[4] Department of Biomedical Informatics,
Stony Brook University, Stony Brook, NY 11794, USA
fusheng.wang@stonybrook.edu
[5] HP Labs, Palo Alto, CA 94304, USA
[6] Department of Computer Science, University of Brasília, Brasília, DF, Brazil

Abstract. Systematic analysis of high resolution whole slide images enables more effective diagnosis, prognosis and prediction of cancer and other important diseases. Due to the enormous sizes and dimensions of whole slide images, the analysis requires extensive computing resources which are not commonly available. Images have to be divided into smaller regions for processing due to computer memory limitations, which will lead to inaccurate results due to the ignorance of boundary crossing objects. In this paper, we propose a highly scalable and cost effective MapReduce based image analysis framework for whole slide image processing, and provide a cloud based implementation. The framework takes a grid-based overlapping partitioning scheme, and provides parallelization of image segmentation based on MapReduce. It provides graceful handling of boundary objects with a highly efficient spatial indexing based matching method, thus avoiding loss of accuracy due to partitioning. We demonstrate that the system achieves high scalability and is cost-effective – our experiments demonstrate that it costs less than fifteen cents to analyze one image on average using Amazon Elastic MapReduce.

Keywords: Whole slide images · Pathology image analysis · MapReduce · Cloud computing

1 Introduction

High-resolution microscopy imaging is playing an increasingly pivotal role in characterizing biological structures quantitatively, revealing new insights into disease mechanisms, and facilitating the development of novel screening exams

© Springer International Publishing AG 2017
F. Wang et al. (Eds.): DMAH 2016, LNCS 10186, pp. 62–77, 2017.
DOI: 10.1007/978-3-319-57741-8_5

and targeted therapies. With a state-of-the-art scanner, a researcher can capture color images of up to $100\,\mathrm{K} \times 100\,\mathrm{K}$ pixels rapidly. A research project could collect datasets consisting of hundreds or thousands of images. Pathology image analysis segments large number of spatial objects, such as nuclei and blood vessels, from whole slide images, along with many image features from these objects. Extracted spatial objects are represented with their geometric boundaries, and such spatially derived information is used in many analytical queries to support biomedical research and exploration. This emergence of pathology analytical imaging fostered by the advent of cost-effective digital scanners has enabled large-scale quantitative and integrative investigations with high throughput analysis of imaging features and annotations [1–3].

However, while digital scanners can produce high resolution whole slide images rapidly, processing such images are complex and highly computational intensive. For example, nuclear segmentation of a single image may extract millions of nuclei and may take several hours with an ordinary desktop computer. It may take weeks for a typical research study with hundreds of images.

High throughput whole slide image analysis demands extensive computing resources, for example, high performance computing clusters [4] and supercomputers, or grid based computing infrastructure [3]. The former approach is highly expensive and often inaccessible, and the Grid approach suffers from major data transportation cost over the Internet. Meanwhile, to handle large scale whole slide images, a common approach is to partition the image into consecutive fixed sized tiles. Objects intersecting the tile boundaries, however, are often discarded, which leads to loss of accuracy in the results. In particular, two potential issues occur with boundary objects. First, an object might be extracted and processed in multiple separate tiles, resulting in duplicates of objects. Second, larger objects, such as blood vessel, might span multiple tiles and its entire geometry is not obtainable by information contained in a single tile, thus it is necessary to merge partial objects.

On the other hand, cloud computing relies on a distributed and interconnected network of computers such as commodity clusters and Amazon EC2 to provide computation, storage, and resource management capabilities elastically in large scale. MapReduce [5] is a popular computing paradigm for cloud computing, with Hadoop being its most common open source implementation. MapReduce gained its popularity in biomedical research [6,7] due to its transparent parallelism, scalability, fault tolerance, load balancing, and most importantly, simplicity in programming.

In this paper, we present a highly scalable and cost effective MapReduce based high performance image analysis framework for whole slide image processing, using a representative nuclear segmentation algorithm [1]. We take an overlapping partitioning scheme providing tile level parallelism for MapReduce. We propose a spatial indexing based matching method combining overlapping partitioning to amend the boundary-crossing problem due to partitioning effect, which will avoid the problem of loss of accuracy from traditional partitioning approaches. The framework is implemented for both commodity clusters and commercial clouds such as Amazon Elastic MapReduce (EMR).

2   Background

MapReduce [5] is a parallel computing framework invented by Google for large scale data analytics. MapReduce not only provides a simplified programming model convenient for parallel application development, but also provides the infrastructure for distributed processing, task parallelization, and fault tolerance. MapReduce programming model comprises two major phases: (i) a map function diving the input problem into smaller sub-problems identified by keys, and distributing them to workers; and (ii) a reduce function collecting answers of sub-problems and combining them to form the output. MapReduce provides distributed processing of both map and reduce tasks. Apache Hadoop[1] is the open source project of MapReduce, which has been widely used to support diverse applications.

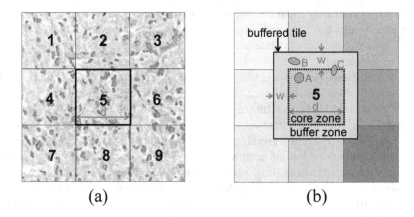

(a)   (b)

Fig. 1. (a) Non-overlapping fixed grid partitioning; (b) Overlapping partitioning

Hadoop Distributed File System (HDFS) is a distributed and scalable file system for the Hadoop framework. It provides large block based storage with distribution replication, offering high availability and fault-tolerance. Besides HDFS on commodity clusters, service based storage such as Amazon Simple Storage Service (Amazon S3)[2] supports HDFS based storage as the backend.

Whole slide images are often represented in a pyramid structure consisting of multiple images at different resolutions [9]. The baseline image has the highest resolution. Due to their substantial dimensions, analyzing and viewing whole slide images are often constrained by computer memory or screen size. Thus, a whole slide image is usually partitioned into much smaller sized tiled images such that a tile can fit into computer memory for image analysis or visualization.

While partitioned tiles can be easily combined for visualization, simple partitioning of tiles will cause problems as objects crossing boundaries will be cut into

---

[1] http://hadoop.apache.org.
[2] http://aws.amazon.com/s3/.

fragments and distributed into multiple tiles as "partial objects" (Fig. 1(a)). Segmentation in such tiles will generate inaccurate results differing from the actual shapes. Most traditional methods simply ignore and remove all such partial objects from the results. For larger and more complex objects such as blood vessels, simple partitioning with ignorance of partial objects will result in the loss of significant results. Thus, a new divide-and-conquer method is needed to support scalable analysis of pathology images with gracefully handling of boundary objects without loss of accuracy or missing of results.

3   Overview of Methods

We propose an overlapping partitioning method for dividing whole slide images into tiles with extended buffer zones (Fig. 1(b)). The core zones are generated from fixed grid partitioning which partitions the whole space as equal sized core zones. Each core zone is enlarged with a buffer so that the majority of objects crossing boundaries of the core zone (boundary-crossing objects) will be contained as complete objects in the buffered tiles. For example, object C crosses the boundary of the core zone and is contained in the buffer zone.

Image processing is applied to each buffered tile, which produces three types of objects: (1) objects fully contained in the core zones, (2) boundary-crossing objects fully contained in the buffer tiles, and (3) buffer-crossing objects which are not fully contained in the buffer tiles. Case 3 can be avoided if the size of the buffer is larger than the estimated maximal dimension (i.e., maximal diameter) of objects. As a boundary-crossing object will be replicated in multiple buffer zones, the image processing will produce duplicated results. We apply a highly efficient spatial-matching algorithm built on top of spatial indexes to remove all duplicated objects (case 2) generated from buffer zones. For case 3, we can apply a spatial stitching algorithm to combine the fragments of an object to produce the whole object through spatial matching of the objects (touching or overlapping) with similar spatial indexing method.

Segmented objects are normally initially produced as image masks, which are converted into geometric based representations such as polygons. The spatial matching will be based on the geometric objects. W present a scalable spatial indexing algorithm using an R*-Tree based spatial indexing based on the *minimal boundary rectangles* (MBRs) of the polygons to merge polygons efficiently [10].

The workflow of the image processing pipeline is shown in Fig. 2, consisting of three major phases: overlapping partitioning to generated buffered tiles, tile based image analysis with MapReduce, and result aggregation with MapReduce. First, original whole slide images are staged onto a distributed file system such as HDFS or Amazon S3. Once the image analysis is initiated, images are read into local disks of each node, where overlapping partitioning is performed to generate tiles. Tiles are copied onto distributed file system for following image processing. Image processing consists of following steps: segmentation to generate mask based results, boundary vectorization to convert masks into polygons, boundary normalization to normalize invalid polygons such as self-crossing polygons. Isolated polygons in the buffer zone which are not crossing the boundary

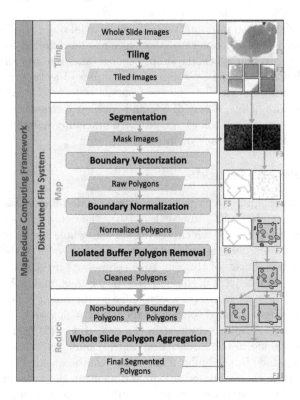

Fig. 2. Overview of the system workflow

of core zones will be removed. Polygon aggregation will merge all partial objects and remove duplicates to generate final results. All steps are parallelized.

4   Overlapping Partitioning

The overlapping partitioning strategy (Fig. 1(b)) generates buffered tiles and each tile consists of a core zone and a buffer zone enlarged from basic fixed grid partitioning (Fig. 1(a)). Buffered tiles are squares with fixed size. The width and height of a core zone are denoted as d, and the span of the buffer zone in each direction is denoted as w. The span of the buffer zone w is determined by the upper bound dimension of segmented objects, such as the maximal diameter of nuclei. This guarantees that every boundary-crossing object of the core zone (e.g., C) will not intersect with the exterior of the buffer zone, thus the object will be segmented as a complete object instead of a partial object from the buffered tile.

Ideally, a minimal buffer width is preferred to minimize the number of objects to be post-processed. However, there exist cases that the upper bound dimension of objects cannot be defined or estimated. For example, blood vessels (Fig. 3) have large extent and the dimensions of segments in each tile could be arbitrary.

For such case, we set a default overlap width w to be 1% of the core zone width d. Boundary-crossing objects that are not entirely contained in the buffer tile will be merged to generate the whole object, discussed in Sect. 6. The width allows spatial index matching method to merge partial objects with overlap accurately.

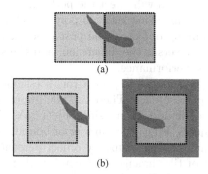

Fig. 3. Example of splitting and merging of blood vessels

Tiling of whole slide images themselves could be parallelized. Whole slide images have a special format that supports random access of a region within an image, without scanning the whole line of pixels [9]. The images are staged onto a distributed file system (HDFS or S3) with a large block based storage, for example, 64 MB as the default storage unit, compared to a typical 4 KB page size in traditional file system. Thus, random access is not possible.

With our approach, instead, each image is fetched onto the local file system of a physical node, where random access is possible to extract regions. We implement a multi-threading based approach to parallelize the generation of tiles. The parallel raster data generation takes advantage of the IO random access, a built-in feature in most vendor-specific formats of Whole Slide Images.

The generated tiles are then copied back to be staged at the distributed file system. The size of each buffered tile is chosen to produce tiles with file size approximately the same as the HDFS block size, e.g., 64 MB, as small files will generate much overhead and slow down the retrieval process.

The distributed file system enables efficient and reliable storage of the tiles, as the data is replicated for redundancy and fault tolerance. The file name of a tile combines the image ID and its top-left coordinates, and forms a unique key for MapReduce based processing, e.g., *TCGA-27-1836-01Z-DX2-0000004096-0000016384*. The tiling speed is directly contrained by the hardware I/O speed and network speed to write tile information into HDFS.

Each generated file contains the relative coordinate of the base corner and the image id, whose combination creates a unique tile id. The information is used as metadata to check for boundary-crossing objects and their respective handling in the merge and duplication removal step. An inherent advantage of staging data on HDFS is fault-tolerance, since each tile might be replicated over multiple nodes.

During processing, if a node fails, MapReduce will fetch a replacement copy of the tile data and automatically start a new job without stopping the workflow.

4.1 Optimal Tile Dimension

We discover that tile dimension will affect the performance. Large tiles could be inefficient as many algorithms' complexity is polynomial. Smaller tiles will lead to higher ratio of overlapping objects to be post processed. Our study shows a parabola like curve of cost versus tile dimension, and the size 4350 is about on the bottom with optimal performance.

We provide a simplified model to give a rough estimate of the cost model for measuring runtime performance. There are two important observation with respect to the number of boundary crossing objects. Assume a relatively uniform distribution of objects. First, the number of boundary objects is linear to the number of partition; thus the boundary object percentage is inversely proportional to the square of tile dimension. Second, with fixed core width d, the boundary object percentage is linearly to the square of buffer width w. These relationships can be proved by statistically analyze the ratio of buffer areas and the total area of the space.

For each image, denote the time to execute each map task as T_1, and the time to execute the reduce task aggregating all boundary objects as T_2, the total number of objects in an image M. Then the total relative run time can be estimated as a function of number of partitions k for each image:

$$
\begin{aligned}
T(k) &= (\sum_{i=1}^{k} T_1) + T_2 \\
&= (\sum_{i=1}^{k}(\alpha + \beta(\frac{M}{k})^c)) + \gamma(M\delta k) * log(M\delta k) \\
&= k\alpha + \beta\frac{M^c}{k^{c-1}} + k\gamma M\delta * log(M\delta) + klog(k)\gamma\delta M \\
&= k(\alpha + \gamma M\delta * log(M\delta)) + k^{1-c}\beta M^c + klog(k)\gamma\delta M
\end{aligned}
\tag{1}
$$

where, α is the relative time overhead to start a mapper job. β and c are coefficients represent the time complexity of the segmentation method and other corrections performed in the map task. γ is the coefficient representing the complexity of the aggregation based on the Algorithm 1, and δ is the ratio number of boundary objects on the total number of objects given the linearity explained above.

Substituting the core width $d = \sqrt{\frac{S}{k}}$ given the area of space S the total run on a single core is:

$$
\begin{aligned}
T(d) &= d^{-2}S(\alpha + \gamma M\delta * log(M\delta)) + d^{2c-2}\beta M^c S^{1-c} \\
&\quad + d^{-2}Slog(S)\gamma\delta M - 2d^{-2}log(d)Slog(S)\gamma\delta M
\end{aligned}
\tag{2}
$$

Observe that the first, third, fourth terms are monotonically decreasing function on core width d, the second term is a monotonically increasing function with respect to d. To obtain the ideal tile size from the equation above, we could take the derivative and find the argument k satisfying the global minimum. Due to the scope of this paper, we omit the empirical method on how to obtain coefficient factors such as α, β, and etc. We plan to explore their behavior more analytically and how to compute the critical value quickly in future works.

5 MapReduce Based Image Segmentation

The image analysis pipeline has two major phases: (1) tile based object segmentation to generate vector represented polygons; and (2) aggregation of tile based results by removing duplicate objects crossing boundaries (Fig. 2). The two phases fit nicely into map and reduce phases of MapReduce.

The first phase is based on titles. For each title, the following steps are performed: (a) Object segmentation to generate results as mask images; (b) boundary vectorization to convert contours of nuclei from mask images into raw "polygons" which might be represented as non-valid polygons; (c) normalize boundaries into valid polygons; and (d) remove isolated objects in the buffer zone which do not intersect core zones.

Object Segmentation. Each tile will be processed with an object segmentation algorithm. Here we take a nuclear segmentation algorithm [1,2] as an example, but the algorithm can be replaced with any segmentation algorithm. This makes the framework highly extensible for other applications. The algorithm includes following steps: preliminary region partition, background normalization, nuclear boundary refinement, and nuclei separation. The results will be represented as black and white mask images, where segmented objects are white filled (F3 in Fig. 2).

Boundary Vectorization. This component is performed in the map task of a MapReduce job. Each tile is directed into a map task, where the key-value pair is the *tile-id* and raster data of the tile, respectively. Other meta data can be provided by appending them to the key. In our study of glioma, we used the image analysis toolkit by Kong et al. [1], which uses morphological operations to normalize the image background, following by applying thresholding methods to segment nuclei. Overlapping nuclei within objects are further separated by watershed techniques. As segmented nuclei will be managed, queried and analyzed, they will be converted into vector based polygons from mask images. Contours of nuclei in the mask images will be extracted as point represented boundaries. Here a function *findcontours* in OpenCV[3] is used to extract contours and represent them as polygons consisting of points.

Boundary Normalization. The machine generated boundaries, however, often contain geometric shapes, which may not be valid polygons based on

---

[3] http://opencv.org.

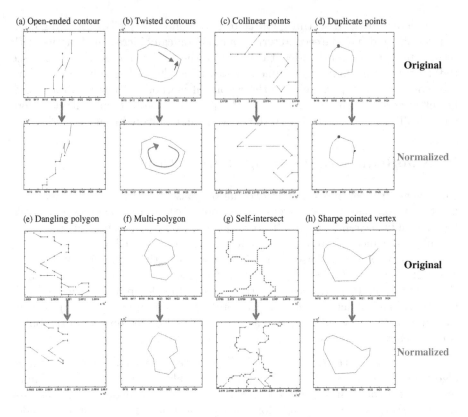

Fig. 4. Examples of boundary normalization

Open Geospatial Consortium standards definitions[4]. These invalid polygons will have vague semantics and prevent further spatial computations.

For example, the shapes may not be topologically closed (e.g., F5 in Fig. 2). We develop a boundary fixing tool to convert boundaries into valid polygons by fixing the following issues: open-ended contour, twisted contours, collinear points, duplicate points, dangling lines, multi-polygon, self-intersect, and sharp pointed vertex (Fig. 4).

– Open-ended contour (Fig. 4(a)): Contours where starting and end points are not identical. A straight-forward correction is to connect the starting and ending points provided there are no conflicting segments like self-intersect. Otherwise, it will be further handled as a self-intersect scenario.
– Twisted contours (Fig. 4(b)): Contours whose directions are changing although the shapes are correct, and it is impossible to iterate linearly on every line segment. This can be fixed by changing the order of the points for contours.
– Collinear points (Fig. 4(c)): Outlier points that have little effect on the overall appearance and structure of a given polygon, which is a result of joining

---

[4] http://www.opengeospatial.org/standards/sfs.

adjacent line segments (Fig. 3b). These outlier points could be removed by reducing and simplifying adjacent segments around these points.

- Duplicate points (Fig. 4(d)): Redundant points. These can be simply removed.
- Dangling (Fig. 4(e)): Extended line segments or a smaller contour connected to the main polygon contour. Dangling segments or contours can be fixed by splitting the contour into parts determined by the presence of intersection, and each part is tested whether it satisfies the properties of a basic polygon; invalid ones are removed.
- Multi-polygon (Fig. 4(f)): Compounded polygons joined by either overlapping or tangent boundaries. Multi-polygon can be considered as a special dangling case, except that individual components form a polygon with distinct shape and area. Thus, this can be handled similarly as the dangling case.
- Self-intersect (Fig. 4(g)): A contour with an edge (line segment) touching or crossing another edge of the contour. This can be resolved by eliminating the exterior coordinates originating from the identified intersection point.
- Sharp pointed vertex (Fig. 4(h)): An angle formed by two-line segments is too narrow. To fix this, the smoothness of the contour has to be refined, for example, by defining a maximal distance parameter on how a point can be away from the main contour.

The boundary normalization tool consists of a pipeline of components from existing libraries, including Boost[5] and Clipper[6] as well as implemented from our custom codes. The boundary fixing tool is robust and fixes all generated boundaries.

Isolated Object Removal. For polygons entirely contained in the buffer zone but not in the core zone, they will not be considered as boundary-crossing objects, thus will be removed. By comparing MBRs of polygons with the buffer zone – representing as four rectangles, the polygons contained in the MBRs will be identified and removed, for example, F8 in Fig. 2.

MapReduce Based Processing of Tiles. The MapReduce framework processes data based on <key, value> pairs. The input to a job is viewed as a set of <key, value> pairs to be processed by the Map function, and a set of <key, value> pairs will be produced as the output of the job by a Reduce function. The <key, value> pairs for Map and Reduce could be of different types, and there will be a shuffle phase between Map and Reduce stages which will sort and regroup the data. Processing of <key, value> pairs in Map and Reduce will be grouped into tasks which will run in parallel and scheduled by MapReduce.

The buffered tiles generated from partitioning form a natural unit for MapReduce based parallel based processing. As indicated by aforementioned steps, each tile is independently processed without having to exchange any information. The tile based image processing will be implemented as a Map function, where the input is a key-value pair of *(tile-id, tileimage)*. MapReduce will scan the input

---

[5] http://www.boost.org.
[6] http://www.angusj.com/delphi/clipper.php.

collection of all tiled images, and each tile image becomes a unit to process by the Map function, which will be executed in parallel as MapReduce tasks running on a cluster. The output of Map function contains records of *tile-id, polygon*, where the tile-id encodes the image ID and the top-left coordinates of the tile. To distinguish polygons intersecting with the exterior boundary of the core zone, an addition attribute is appended to the key *tile-id* to label it. Only labeled polygons will be merged in the next Reduce step.

6   MapReduce Based Spatial Processing and Result Aggregation

The image analysis will produce polygon represented contours of segmented objects. As the execution is performed on each title independently, boundary-crossing objects will be produced from multiple tiles in the intermediate results from Map function.

Polygons from previous step will fall into two categories: non-boundary-crossing polygons, for example, polygons in F9 (Fig. 2), and boundary-crossing polygons, for example, polygons in F10. Boundary polygons are labeled by appending a tag in tile-id. The non-boundary-crossing polygons will be output directly and no additional processing is needed. Boundary-crossing polygons will be combined, matched and merged in the Reduce phase.

In Reducer phase, a new key is formed which is the image ID, and the shuffling will be performed by grouping all polygons of the same image. As a result, all duplicate objects will be combined together for an image. We take advantage of the sorting and shuffling mechanism of MapReduce which is highly scalable. For the collection of objects for each image (value of an image ID key), spatial matching and merging is performed through a spatial indexing based approach. The Reducer will parallelize all the processing based on image ID key.

6.1   Object Merging Algorithm

We propose an spatial indexing based object merging algorithm (Fig. 1). For all the labelled boundary objects of the same image, an R*-Tree is built based on the MBRs of the polygons [10]. The R*-Tree will be stored in memory as the percentage of boundary objects is usually small, for example, a few percent. R*-Tree allows extremely efficient search of intersecting objects in the image.

As polygons with intersecting MBRs will be grouped together based on the spatial locality in the R*-Tree, it will be highly efficient to identity the duplicate objects or perform object-merging. For an image with larger number of boundary objects, we apply a bottom-up approach grouping neighboring tiles and constructing an R*-tree for each region. We could successively apply the merge strategy until there are no boundary objects. In our implementation, we take advantage of the transitivity nature between partial objects: while maintaining new merged objects in memory, we only need a single pass over each object which incurs a total of $O(nlog(n))$ time for every case, where n is the number of objects to be merged.

Algorithm 1. Partial Objects Handling

```
1  vectorObjects = getInput() ;
2  finalObjects = getEmptyList() ;
3  rtreeindex = buildRtreeIndex(vectorObjects) ;
4  key = 0 ;
5  for object o in vectorObjects do
6      if o.id == -1 then
7          o.id = key ;
8          finalObjects.insertAt(o, key) ;
9      end
10     intersectingNeighbors = rtreeindex.findIntersection(o);
11     for object t in intersectingNeighbors do
12         if t.id ≤ o.id or o.id == −1 then
13             skip ;
14         end
15         prevMerged = finalObject.getByPosition(o.id) ;
16         currentMerged = union(prevMerged, t) ;
17         finalObject.replaceAtPosition(currentMerged, o.id) ;
18         t.id = o.id ;
19     end
20     key = key + 1;
21 end
22 output(finalObjects)
```

The algorithm uses parallel arrays to store geometry information and the id of the location the logical merged object belongs to. Objects are sequentially stored in an array list, where each id is the position of the object in the array. We take advantage of the property that every partial object belonging to the same object has to intersect with at least one another object. As we process an individual object, we apply intersection search using R*-Tree to obtain intersecting neighbors. The geometries of these partial objects will be merged into the existing object identified by the first original partial object id.

In other words, all objects logically belonging to the same object will contain partial id of the first object (i.e., object with the lowest id). This id value will be propagated down the list and updated to other objects intersecting with any existing object from the same set of partial objects. In addition, we track and update the total final merged object geometry by storing its location to the pointer of the first object from each partial object set.

Algorithm complexity. The spatial merging algorithm has a direct complexity of $O(n\log(n))$, where n is the total number of boundary objects to be processed. Furthermore, due to the linear properties of image independent parallel processing, the algorithm has a $O(m\log(m))$, where m is the number of objects in all image.

The final object mark-ups are stored on HDFS, where they can be used for future studies and analyses.

7   Performance and Evaluation

Experiments Setup. We use a data set with 475 whole slide images of brain tumor from TCGA portal[7], with size of 1.3 TB. The data set contains more than 180 million nuclei to be extracted. To test the object merging and duplicate removal algorithm, we use blood vessel data synthetically generated to simulate a wide range of object complexity. We use an Amazon EMR cluster consisting of up to 200 c3.large nodes, each equipped with 3.75 GB of RAM, and a Intel Xeon E5-2680 2.8 GHz processor with two cores. The maximal number of cores is 400.

Performance of Tiling. The whole slide images are first staged to Amazon S3. Images are first tiled, and buffered tiles have a dimension of 4350×4350, with buffering width 127 and core zone dimension 4096×4096. The tiling is performed in parallel on Amazon EC2 cluster, and the generated tiles are copied from local disks back to S3. We only use 10 nodes to perform tiling as it is sufficient to generate tiles quickly. We find using multi-threading on each node will increase the tiling speed significantly. As shown in Fig. 5(a), the tiling time for all images drops from 67 s to 10 s when the number of threads is increased from 1 to 16. This is due to the fact that I/O is the bottleneck of tiling, but SSD could deliver high performance reading with multi-thread reading [11].

Scalability of MapReduce Based Image Processing. Figure 5(b) shows the performance of MapReduce based image processing, with increasing number of CPU cores, which demonstrates almost linear scalability. By increasing cores from 32 to 400 (a factor of 12.5), the time drops from 1045 min to 90 min (a factor of 11.6). It is projected that without parallelization, it will take more than 23 days to finish the processing. Faster processing can be easily achieved through increasing the number of cores.

Figures 5(c) and (d) demonstrate the performance of the polygon merging algorithm per image in MapReduce. Figure 5(c) shows merging time on blood vessel data of different complexity. While the number of total unique blood vessels is fixed as 10 per images, the number of fragments generated by individual tiles are variable. We observe that with increasing complexity, the merging is becoming quadratic due to the increasing number of vertices and edges, which significantly affect the geometrical computation. On the other hand, Fig. 5(d) shows the merging time on a dynamic number of unique blood vessels with a total fixed number of 4,000 fragments per image. We observe that the merge time is inversely proportional to the number of blood vessels. This implies that the merging time is approximate quadratic to blood vessel complexity and is linear to the number of blood vessels.

---

[7] https://tcga-data.nci.nih.gov/.

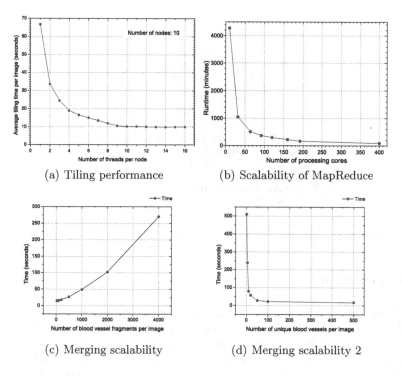

(a) Tiling performance   (b) Scalability of MapReduce

(c) Merging scalability   (d) Merging scalability 2

Fig. 5. Tiling and overall runtime performance

Profiling of the Image Processing Pipeline. We also break down the execution time of the image processing pipeline (Fig. 6). Data staging takes 8.7%, tiling takes 13.1%, segmentation takes 57.6%, vectorization and normalization takes 10%, and aggregation takes 10.5%. The merging of duplicated polygons in the aggregation step is only a small fraction of the total cost due to the effective spatial indexing based approach.

Cost Estimation. We also evaluate the cost for processing images on Amazon cloud. Amazon EC2 has different node types with different pricing. For example, c3.large comes with 2 cores and costs $0.105 per hour, and c3.8large comes with 32 cores and costs $1.68 per hour. We take c3.large as it is more cost-effective. We only use 10 nodes for tiling as it is sufficient to produce tiles quickly, and we use 200 nodes with 400 cores for MapReduce based image processing. The total cost is approximately $36.25, including $27 for using EMR for image processing and $9.25 for tiling. Amazon S3 storage is about $0.0300 per GB per month. As all the uploading, processing and downloading could be finished within 2 h, it will take less than two hours to store the images and tiles. The storage cost is about 20 cents and can be ignored. The data uploading is free, and data downloading costs $0.12 per GB. The result is about 100GB, which costs $12 to download. Thus, the total cost will be about $48, and the average cost per image is about 10 cents.

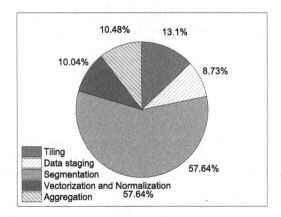

Fig. 6. Execution time breakdown of the image processing pipeline

8 Conclusions

MapReduce based programming model and cloud computing becomes a de facto standard for large scale data analytics due to the scalability, fault tolerance, simplicity and cost-effective operations. In this paper, we study the feasibility of applying the software stack for large scale whole slide image analysis, by solving the critical issues of partitioning, inaccuracy, and parallelization. We propose a generic MapReduce based workflow with tile level parallelism, by developing a buffered overlapping partitioning method supported with a spatial matching based normalization method for amending boundary effect. The method provides accurate results with a very small overhead with efficient indexing based matching. Parallelization is thus achieved transparently through MapReduce, which achieves almost linear scalability. Our experiments demonstrate high throughput and low cost of the approach: for analyzing 475 whole slide images, it can reduce computation time from 23 days to less than one hour on Amazon EC2 with an average cost of 10 cents per image.

Acknowledgements. This work is supported in part by NSF IIS 1350885, by NSF ACI 1350885, by Grant Number K25CA181503 from the National Institute of Health, by Grant Number R01LM009239 from the National Library of Medicine, by Grant Number 1U24CA180924-01A1 from the National Cancer Institute, and by CNPq.

References

1. Kong, J., Cooper, L.A.D., Wang, F., Teodoro, G., Scarpace, L., Mikkelsen, T., Schniederjan, M.J., Moreno, S., Saltz, J.H., Brat, D.J.: Machine-based morphologic analysis of glioblastoma using whole-slide pathology images uncovers clinically relevant molecular correlates. PLoS One **8**(11), e81049 (2013)

2. Cooper, L.A.D., Kong, J., Gutman, D.A., Wang, F., Gao, J., Appin, C., Cholleti, S., Pan, T., Sharma, A., Scarpace, L., Mikkelsen, T., Kurc, T.M., Moreno, C., Brat, D.J., Saltz, J.H.: Integrated morphologic analysis for the identification and characterization of disease subtypes. J. Am. Med. Inform. Assoc. **19**(2), 317–323 (2012)
3. Foran, D.J., Yang, L., Hu, J., Goodell, L.A., Reise, M., Wang, F., Kurc, T., Pan, T., Sharma, A., Saltz, H.: Imageminer: a software system for comparative analysis of tissue microarrays using content-based image retrieval, high-performance computing, and grid technology. JAMIA **18**(4), 403–415 (2011)
4. Teodoro, G., Pan, T., Kurc, T.M., Kong, J., Cooper, L.A.D., Podhorszki, N., Klasky, S., Saltz, J.H.: High-throughput analysis of large microscopy image datasets on cpu-gpu cluster platforms. In: IPDPS, pp. 103–114, May 2013
5. Dean, J., Ghemawat, S.: MapReduce: simplified data processing on large clusters. Commun. ACM **51**(1), 107–113 (2008)
6. Aji, A., Wang, F., Saltz, J.H.: Towards building a high performance spatial query system for large scale medical imaging data. In: SIGSPATIAL GIS, pp. 309–318. ACM (2012)
7. Aji, A., Wang, F., Vo, H., Lee, R., Liu, Q., Zhang, X., Saltz, J.H.: Hadoop-GIS: a high performance spatial data warehousing system over MapReduce. Proc. VLDB Endow. **6**(11), 1009–1020 (2013)
8. Cooper, L.A.D., Kong, J., Wang, F., Saltz, K.T., J.H., Brat D.: In silico analysis of nuclei in glioblastoma using large-scale microscopy images improves prediction of treatment response. In: EMBC (2011)
9. Wang, F., Oh, T.W., Vergara-Nidermayr, C., Kurc, T.M., Saltz, J.H.: Managing and querying whole slide images. In: SPIE Medical Imaging (2012)
10. Beckmann, N., Kriegel, H., Schneider, R., Seeger, B.: The r*-tree: an efficient and robust access method for points and rectangles. In: SIGMOD (1990)
11. Zhang, X., Wang, F., Lee, R., Saltz, J.H.: Towards building high performance medical image management system for clinical trials. In: SPIE Medical Imaging, pp. 762805–11 (2011)

Information Extraction and Data Integration for Biomedical Data

Drug Dosage Balancing Using Large Scale Multi-omics Datasets

Alokkumar Jha[✉], Muntazir Mehdi, Yasar Khan, Qaiser Mehmood,
Dietrich Rebholz-Schuhmann, and Ratnesh Sahay

Insight Centre for Data Analytics, National University of Ireland, Galway, Ireland
alokkumar.jha@insight-centre.org

Abstract. Cancer is a disease of biological and cell cycle processes, driven by dosage of the limited set of drugs, resistance, mutations, and side effects. The identification of such limited set of drugs and their targets, pathways, and effects based on large scale multi-omics, multi-dimensional datasets is one of key challenging tasks in data-driven cancer genomics. This paper demonstrates the use of public databases associated with Drug-Target(Gene/Protein)-Disease to dissect the in-depth analysis of approved cancer drugs, their genetic associations, their pathways to establish a dosage balancing mechanism. This paper will also help to understand cancer as a disease associated pathways and effect of drug treatment on the cancer cells. We employ the Semantic Web approach to provide an integrated knowledge discovery process and the network of integrated datasets. The approach is employed to sustain the biological questions involving (1) Associated drugs and their omics signature, (2) Identification of gene association with integrated Drug-Target databases (3) Mutations, variants, and alterations from these targets (4) Their PPI Interactions and associated oncogenic pathways (5) Associated biological process aligned with these mutations and pathways to identify IC-50 level of each drug along-with adverse events and alternate indications. In principal this large semantically integrated database of around 30 databases will serve as Semantic Linked Association Prediction in drug discovery to explore and expand the dosage balancing and drug re-purposing.

1 Introduction

Drug discovery and dosage balancing in cancer have always been influenced by biological studies and clinical parameters. However, rapidly growing genomics studies and the gambit of data produced by next generation technology in last decade opened up a window to understand the drug delivery in cancer based on a mechanism and not just only observations [24]. The biggest handicap in this process is the identification of datasets in this multi-layer study, rank them based on their curation level and link them based on their instances to avoid false-positives. There are certain challenges with this study, such as cancer has very limited set of medicines and combinations of these in different dosage is to be

© Springer International Publishing AG 2017
F. Wang et al. (Eds.): DMAH 2016, LNCS 10186, pp. 81–100, 2017.
DOI: 10.1007/978-3-319-57741-8_6

used to target more than 30 cancer types. The detailed study of the most effective anti-cancer drugs against various tissue types have been cataloged categorically in cancerDR [1].

The key problems in cancer drug discovery and establishing an integrated Drug-Target-Disease(DTD) association is to identify exact causative factors that contribute semantics in the disease, such as identification of drug targets, Mutations data, and Drug-gable structures. In the present paper, we have introduced a novel approach for drug discovery keeping most effective drugs in cancer as a seed to query each layer of linked databases grouped together based on similarities to achieve enriched annotation for each entity. Once the information has been retrieved to achieve the highlighted 5-point agenda, further linking has been done to establish an end to end pipeline from drug to dosage distribution based on omics dataset. This have been done by tracking all possible activities and effects that could result by a drug or compound. Figure 1 gives a brief insight into the proposed approach, each layer of the Fig. 1 is further detailed in the coming sections of the paper. We have identified that major biological processes involved with this drug and associated diseases are Apoptosis which explain its connections.

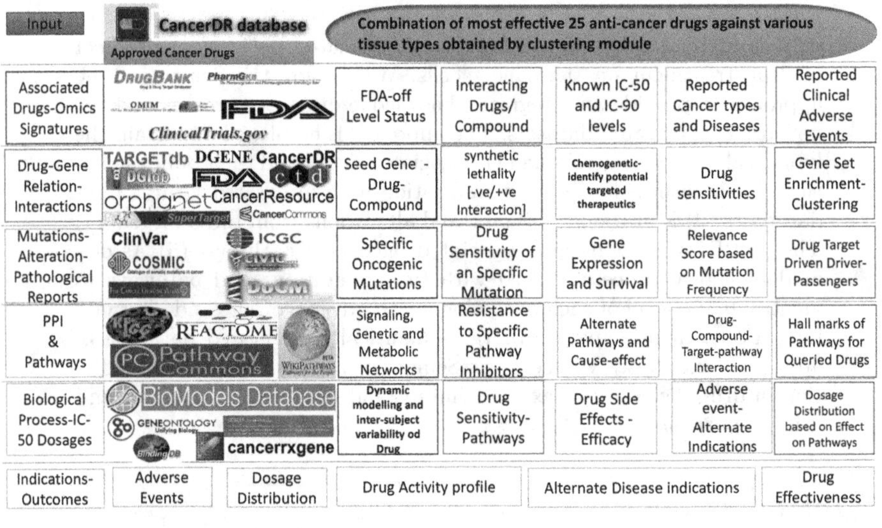

Fig. 1. Multi-layer database driven knowledge discovery platform (Color figure online)

As explained in Fig. 1, we have divided the databases to solve the problem based on similar source-enriched outcome strategy. The iteration works in such a way that in the first layer we will identify the entities associated with 25 cancer Approved Drugs. To formalize the problem of drug distribution we have modeled the whole problem in subway architecture as shown in Fig. 1. The green bar in the figure explains the known entities for already approved drugs called as known-subway stations. And the red bar explains the stations which need to be identified to find the alternate path and distribute the station load (SPARQL endpoints) with similar sources to have least congestion path called as alternate indications from Drug. These least congestion path stations have been identified as a combination at each stage in the form of databases of each stage.

The first layer is *Source Station*, where we identify the drug, associated compounds, approval status, interacting drugs, approved dosages effects and other disease or cancer types. The second layer, which we call *Terminal-1 (long-route departure point)* because this stage is used in the sources specified in Fig. 1 to help extract further drugs, orphan drugs, associated genes, gene set enrichment and confidence level of each Target(gene/protein). And because this stage will enrich set of gene list which will be used to understand the other genetic aspects of drugs such as mutations, methylation etc. The Third layer is *domestic departure point* where the set of databases from Fig. 1 will provide the known mutations, variants, Gene Expression(GE) and survival information for extracted gene list in second layer. In cancer, identified mutations can be modeled for according to desired drug target and have short term impact, this stage names as the *domestic departure point*. The next layer named as *Junctions*, where the junction represent pathways and interacting partners of the seed genes from approved drugs in cancer. These Pathways have already been modeled as subway model [3]. This layer will try to identify involved processes from this existing subway model by drug dosage balancing. Higher or lower drug dosage in cancer leads to select alternate pathways in cancer [4].

Our approach has identified these changes of pathways and cancer cell processes to understand the alternate indications from Drug. The last station and stage-5 in Fig. 1 represents the arrival station. This station will take care of maintenance and tools required in terms of maintenance such as GO biological processes and repairing genes, adverse events, side effects, and drug-dosage tolerance. This five subways station subway problem algorithm have been described as a solution for 25 FDA-approved anti-cancer drugs in coming sections of the paper. The flow of the paper is as follows: Section 2 will explain the subways and drug discovery problem in detail as a motivational scenario. Section 3 will explain the related work in terms of biological solutions and algorithm solutions. Section 4 will explain the discovery and linking mechanism, and gives a brief insight into the data complexity statistics and other challenges. Section 5 will explain the solution of each station based on algorithm explained in Sect. 4 with four queries, their results and significance of drug discovery. Finally, we conclude our work on the findings for these 25 anti-cancer drugs.

2   Motivational Scenario

One of the key learnings in cancer is to identify new drugs based on the genomics signature of current drug or balance of the dosage to find alternative treatments. In principal, it is a mechanism based effort beginning from approved cancer drug to associated cancer gene discovery effort that intends to discover the full set of

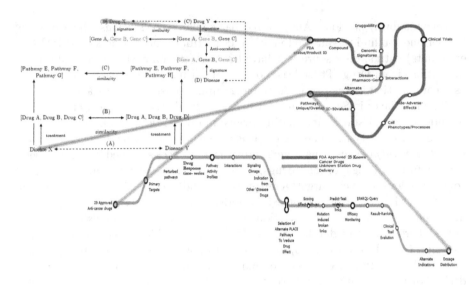

Fig. 2. Algorithm driven subway pipeline model (Color figure online)

amplified oncogenes that play primary role in causing other cancers. We have extended the Highway approach [2] to linked data based multi-layer linking subway approach for dissecting the mechanism of each approved drug involved in cancer. Figures 1 and 2 represents the key motivations to dig deeper into the problem. Figure 1 explains the data-driven challenges which can in parallel be associated with Fig. 2 for a biological problem and links associated with drug discovery. We have modeled this pipeline as subway issue in Fig. 1 and prospective drug discovery pipeline mapped with each database at each stage as Fig. 2[1,2]. The bottom red subway, Fig. 1 will be annotated with Fig. 2 to identify the key problem associated with cancer drug discovery and dosage balancing. To understand the associated challenges, lets assume one of the 25 anti-cancer drug from kumar et al. [1]. Detailed results will be discussed in Sect. 5. One such Drug is *Paclitaxel*, which has been reported with multiple tissues such as Autonomic Ganglia, Bladder, Blood, CNS, Endometrium,Lung, Oesophagus, Ovary, Pancreas, Pleura, Prostate, Salivary Gland, Skin, Soft Tissue, Stomach, UAT* and Urinary Tract. Being the most common anti-cancer drug, it has been associated with almost all types of primary tissues in cancer. Which leads then to be converted into 3-times of primary tissues in carcinoma. Now drug distribution among these tissue type will depend on the association of pathways with these drugs [9].

To Understand the dynamics of this drug, we have to dig data as per the 5-layer strategy, mentioned in Fig. 1. For the first layer, we have excluded the tissue list and cancer types associated with this drug within this layer. We ran

---

[1] Figure 1 is partially adapted from [5].
[2] Figure 2 is remodelled using [3,6–8].

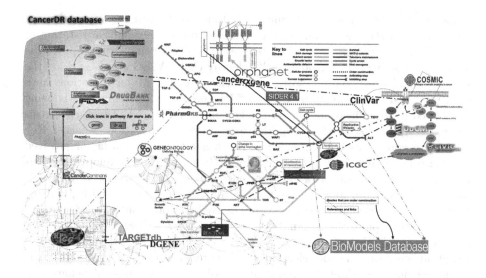

Fig. 3. Domain driven subway pipeline model

an integrated query to obtain existing knowledge of this drug from Drug Bank, PharmaGKB, ClinicalTrails, FDA, and OMIM. We have re-used Bio2rdf [28] and developed dedicated SPARQL endpoint for each layer and linking of databases having been performed as per Sect. 4. This layer will give us monotherapy data consisted of estimated drug response curve parameters such as IC50 and AUC estimates. As explained in Fig. 2 this drug is majorly active in the breast cancer, ovarian cancer and the genes highly expressed in this region are FASLG, TNFSF6, APT1LG1, FASL, ALPS1B, EGFR. However, Endpoint-1 also provides the key information such as pathways associated with this drug and status of the clinical trail. We will use the key learning from Endpoint-1 to dig deeper from genes highly expressed and the most highly ranked cancer types. This layer shows an ideal case why these data sources need to be linked together for knowledge enrichment and also understand the drug associations. Similarly, this will help to find Druggability subway station as per solution perspective. As mentioned in Fig. 2 we have further looked into genomic signatures subway station Fig. 1, where we have linked databases to explore interacting genes, their associative diseases and orphan diseases, pathways to dissect the mechanism of Drug in cancer since cancer often identified as multiple diseases and orphan diseases play a crucial role in cancer development [10]. This explains its pathway associations as mentioned in Fig. 2 and even on data level, we have obtained the similar pathway connections such as Neurotrophin signaling pathway. The Drug interactions with ERLOTINIB, GEFITINIB, EGFR INHIBITOR, because of other targets such as PTEN, ABCC10, PDCD4, TIMP1 and close combinations with alternate and rare diseases (Type I diabetes, Autoimmune lymphoproliferative syndrome, Neonatal inflammatory skin and bowel disease, Lung, Colon,

Blood, Cell). The Enriched information is tough to be retrieved from the single source and strategic lining of layer two provides a key subways station of look forward towards genomic alteration associated with this Drug to prove the above shortcomings. TCGA and ICGC identify that FASLG and EGFR are the key genes and other markers work as associative genes. Major mutations have been identified in Lung cancer, Uterine, Melanoma and mutation type is being dominated by amplification. One of the key outcomes is TNF(Tumour Necrosis Factor) associated with this drug and responsible for mutation. Another Key outcome is coming as GATA2 receptor which is a transcription factor involved in stem cells maintenance with a key role in hematopoietic development. This is evident that selection of target from one source and integration of source in this layer-3 have determined the mutation-driven cause associated with this drug. However, it also provides and nearest supportive junction to subway model which helps to even explore the assignations shown in Sect. 2 where TNF is indeed a key factor involved in Apoptosis. This approach helps to fill the gap between mutation based driven gene selection and drug resistance. Once we obtained the mutations and variations, it's essential to look into the change of behavior in the cell due to these variants/mutations. Subway Point Pathways helps to find this cause and as mentioned in Fig. 2 this drug associated hsa04010-MAPK signaling pathway, hsa04014 -Ras signaling pathway, hsa04060 -Cytokine-cytokine receptor interaction, hsa04068 -FoxO signaling pathway along with REACTOME signaling by ERBB2, Semaphorin interactions, Acetylcholine regulates insulin secretion, Protein repair. The interaction due to this pathway changes key interactors: MIMAT0000083, MIMAT0000076, RIPK1, FAS, FADD, CHUK, CASP8, CASP3. The key is the involvement of nCounter Prognostic AML miRNA signature which provides pathways associated with side effects attached to this drug. This leads to motivate to visit layer-5 and subway station called as alternate indications where due to this altered pathways we have associated side effects such as Leukopenia, Neutropenia, Alopecia, Anaemia uncommon. The subway path has also involved few of the processes mutations such as GO:Tumor necrosis factor ligand superfamily member 6, Proctor2005 - actions of chaperones and their role in aging. This annotation has opened a window for this drug involvement in aging and the binding site for such association would be taxol CHEMBL428647 BDBM50001839 PACLITAXEL. After all this activity the key questions arises the optimal dosage due these multi pathways associations and involvement in rare and orphan diseases. We have identified the optimal dosage as IC-50 level the, screened-356,Maximum IC50-31.5, Geometric mean-0.0882, Minimum IC50-0.000287. The key motivation behind this research could be well understood from the summary in Fig. 2 where we have dissected the drug from all perspective and it is evident that multi-source linking to connect dots as subway station is key in drug discovery and distribution.

3   Related Work

Present paper covers aspects such as ADR, alternate indications, pathways, orphan diseases, and mutations. All these studies having been done as the

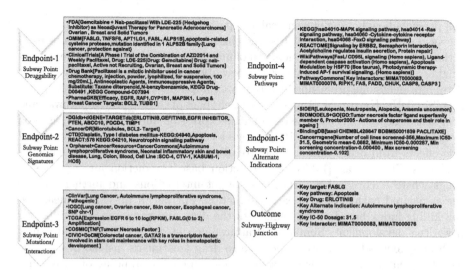

Fig. 4. Domain driven subway pipeline model illustrated using single drug example: Drug repositioning key challenge explained using "paclitaxel" as control and solving strategy using subway station identification approach [Fig. 2] and following layered approach described in [Fig. 2] [Top -2 results only for each]

individual platform either dedicated to a particular database or dedicated disease/cancer type. Croset et al. [5]. In this approach, authors have investigated drug repositioning using the concept of mode of action to identify new indications for approved or failed drugs. This computational approach has explored Drug bank database by identifying drugs perturb multiple biological entities themselves involved in multiple biological processes. Similarly, for clinical trails approach for semantic link discovery have been used to identify the involvement of drugs using state-of-the-art approximate string matching techniques combined with the ontology-based semantic matching of the records. Evaluation of the performance of our proposed techniques in several link discovery scenarios in LinkedCT [11]. [8] explains the impact of adverse events on drug repositioning and drug failure research covers that Adverse drug reactions (ADRs) are responsible for drug candidate failure during clinical trials. It is crucial to investigate biological pathways contributing to ADRs. Large-scale analysis to identify overrepresented ADR-pathway combinations through merging clinical phenotypic data, biological pathway data and drug-target relations would be crucial in cancer drug repositioning as well. From Algorithms perspective [2] presents the gaps in the pathways to understanding the drugs using highway model. This could be correlated with Subway model presented in this paper. Identification of side effects from the drug has been presented based on predict potential side-effects of drug candidate molecules based on their chemical structures, applicable on large molecular databanks [12]. The key message from the present paper is the impact of strategicaly linking multi-dimensional data sources and zheng et al.

[13] presents how linking of chemical and pathways databases can help to obtain the systematic prediction of ADRs, the characterization of novel mechanisms of action for existing drugs. The inclusion of multi-omics data has always been challenging and ma et al. [14] provides a review about pathway analysis of genomics data, represents one promising direction for computational inference of drug targets. Similarly, Karp et al. [15] explains the impact of the integrated pathway and drug database analysis and the role of gene expression and cell survival based on drug treatment. ATIAS et al. [16] presents a comprehensive algorithm to predict side effects from drugs and a model to adhere the changes based on genomics data sets. Zhou et al. [17] is an interaction based prediction and analysis of drugs where it covers drug-drug, drug-target interactions and further analyse the associated pathways with each interaction. Further, it provides a measured call killing index based on side effects of the drug and results in a pre-filter for new drug development. Similarly, Pratanwanich et al. [18] demonstrates how pathways/gene sets that are responsive to drug treatments instead of a simple list of regulated genes. This can also advance our understanding of such cellular processes after perturbations. Apart from the study of adverse events and target from a drug, it's equally essential to demonstrate the drug dosage distribution which in principal is the key message from present paper. The similar approach focused just on the expression level of the genes have been presented as ToXDB [19]. Pathways can individually pay the key role in the drug repositioning and this has been demonstrated using the example of Crohn's Disease with Li et al. [16]. At last, efficacy is one of the key factors in drug discovery and since repurposing has become essential in the era of where clinical trails are tightly regulated and failed approved drugs can play a significant role if analyzed through genomics approach. Guney et al. represent the network based approach which has also influenced few portions of current paper which explain the results [21]. As we can see all related work covered one of the other dedicated approach fro dedicated task in drug discovery. However, there is certainly a requirement for single linked databases platform where all these approaches can be integrated on a systems level with a mechanism as mentioned in Fig. 2. And present paper has tried to achieve that with appropriate efficiency using federated semantic model and linking.

4   Materials and Methods

4.1   Linking

We employ a linked data approach to answer and support the biological questions of this paper. Linked Data is a paradigm for publishing structured data on the Web following four core principles: (1) use URIs to name things, (2) use HTTP URIs so those things can be looked up, (3) provide structured data about those things when their URIs, (4) provide links to related datasets. Taken together, these principles recommend publishing structured data in an interoperable format (RDF), where URIs serve as both names for things and addresses

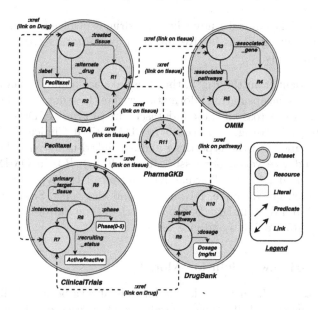

Fig. 5. Linking scenario illustrated

for their description, and where links enable automatic discovery of remote content. Following the fourth Linked Data principle: *"link to related data"*. From the perspective of a consumer, these links allow for recursively discovering and navigating detailed information about related entities among different datasets. These links encourage modularity, where high-quality links (once in place) can reduce the amount of content needed to be hosted.

In our approach, we primarily create a set of query terms using [26] for drugs, pathways, genes or tissues. Afterwards, we use those query terms to discover relevant datasets using the techniques presented in [22,27]. Once the identification of the entities on disperse datasets has been achieved, the next thing is to identify the key information these entities contain. Thus, finally, we create soft links between similar entities dispersed in different datasets. An example linking scenario is given in Fig. 5, where these soft links (:xref) can be seen. The presented example scenario is generated for linking the datasets involved in checking the druggability of approved cancer drugs. The datasets involved in the said scenario are (1) FDA, (2) OMIM, (3) DrugBank, (4) PharmaGKB, and (5) ClinicalTrials. To check the druggability, we require the following information from the aforementioned datasets:

– FDA: Drug, its alternative drug, and the tissue it targets.
– OMIM: Tissue, its associated pathways and genes.
– DrugBank: Drug, the pathways it targets, and the dosage information.
– PharmaGKB: Information on tissue.

– ClinicalTrials: The clinical study where the drug is involved, other intervening drugs, the recruiting status (active/inactive), primary target tissue and phase of the clinical study.

It is further evident from the Fig. 5, that the links are mainly generated by the common resources encapsulating information within these datasets. The common resources dispersed across all these 5 datasets are of type (1) Drug, (2) Tissue, and (3) Pathways. For drug, resource R0 (in FDA), R7 (in ClinicalTrials), and R9 (in DrugBank) are all of the type Drug and are linked using :xref. For example R0 :xref R7 and vice versa. For tissue, R1 (in FDA), R3 (in OMIM), R8 (in ClinicalTrials), and R11 (in PharmaGKB) are all of the type Tissue and are linked using :xref. For example, R1 :xref R3 and vice versa. Similarly, resource R5 (in OMIM) and R10 (in DrugBank) are both of type Pathway and are linked using :xref like R5 :xref R10 and vice versa.

4.2   Subway-Highway Solution Model

Knowledge discovery from multiple databases is always been a challenging job aseptically in the case of life science databases where each database add a new knowledge to the existing knowledge and identification of key links among this is key. We have discussed the linking approach in our distributed data set in Sect. 4.1. However, another key challenge is discovery when data sets can also contain false positives, redundant information, curated and noncurated information. The discovery algorithm must have a ranking method to rank the incoming data based on certain criteria. As discussed in Sect. 2 we have modeled our drug discovery approach using Subway model and we have extended this in such a way where all five layers from Fig. 1 can be independent Subway station and they need to integrate and connected with highway model to obtain desired results. The Highway model can be understood the as best known knowledge discovery platform as we have modeled in Fig. 2. The algorithm has five layers as mentioned in Fig. 1. And below si the description of each layer along with key outcome and solution to find disjoint subways stations and their connecting stations.

Assumptions

- Input drug list of 25 anti-cancer drugs from [1] will be considered as establishing locations going to be connected by multi-layer subway.
- FDA approved cancer-diseases having been considered as established Subway stations in the network.
- FDA approved non-diseases having been considered as stoppage on request stations in the network.
- Any cancer indications from any other database will be considered as waiting for the area of subway stations.
- Any non-cancer indications from any other database will be considered as a general-waiting area of subway stations.

- Signature Targets/Genes for/from cancer will be identified as start signals in the subway.
- Signature Targets/Genes for/from non-cancer will be identified as stop signals in the subway.
- Pathways for/from cancer will be identified as established critical tracks among the subway stations.
- Pathways for/from non-cancer will be identified as under construction and wait for signal subway tracks.
- Pathways for/from both cancer/non-cancer will be identified as waiting for other tracks to be connected(pathways) stage.
- Alternate indication with side effect will be considered as work in progress stop junctions.
- Alternate indication without side effect will be considered as abandoned stations of the subway.
- Cancer IC-50 values will be called as desired destinations.
- Non-Cancer IC-50 values will be called as destinations at equal distance from all major stations.
- Any Orphan or rare disease related to cancer will be called as station overload responsible for pathway malfunction.
- Any Orphan or rare disease related to non-cancer will be called as station overload responsible for shorter path construction.

Based on the assumptions above we have built a subway model to connect all possible stations and find an optimal path among subway stations. The overall methodology is divided in five-step sequential process.

Step-1: Identify hubs-sub-hubs, paths and predicted paths for Subway station[Know and unknow attributes about approved cancer drugs]

Input: Paclitaxel, Vinblastine, Vinorelbine, Panobinostat, 17AAG, Camptothecin, Elescomol, Thapsigargin, EpothiloneB, PD0325901, AZD6244, Nutlin3, Docetaxel, AUY922, Panobinstat, AZD0530, Irinotecan, Topotecan, Vinrelbine, LBW242, Nilotinib, Gemcitabine, Vinorelbin, Doxorubicin, Methotrexate, Topetecan, AEW541, EpothilonB

Output: Alternate Diseases(cancer/non-cancer) & enriched target gene list

As mentioned in Assumptions:

FDA Approved Tumor Type → Main-hub Stations

FDA Approved other disease → disjoint and unidentified hub stations

If(database = OMIM) Target → Genes→Known path to the stations Biological Process → connection to the disjoint station tracks

Tissue-Disease Type→ Compulsory Hub to be build for cell survival

IF(HubStation$_{FDA}$ $\bigcap$ HubStation$_{OMIM}$) = Cancer Tissue

then Result1 = Metro-Junction Hub(Approved Cancer type status)

ELSEIF(ClinicalTrail = recruiting && ClinicalTrail = Cancer)

Update(Result1 = Cancer With Subtypes)

ELSEIF(HubStation$_{FDA}$ $\subset$ HubStation$_{OMIM}$)$\bigcap$HubStation$_{FDA}$ = Cancer Type Break;

So, Result2 = Alternate Cancer type(Hub exit point towards cancer pathways)

Query Result1, Result2 $\rightarrow$ PharmaGKB, OMIM, DrugBank

IF(PharmaGKB$_{\text{Target}}$ $\cup$ DrugBank$_{\text{Target}}$) = FDA Target(Gene/Protien)

Result3 $\rightarrow$ Approved Entry as Sub-Station(Know non cancer drug signature genes)

ELSEIF(PharmaGKB$_{\text{Target}}$ $\cup$ DrugBank$_{\text{Target}}$) = OMIM Target

Update Result3 $\rightarrow$ Expected or Predicted entry to substation(Interacting partner drives to alternate disease)

ELSEIF((PharmaGKB$_{\text{Target}}$ $\cap$ DrugBank$_{\text{Target}}$) $\subset$(FDA$_{\text{Target}}$ $\cap$ OMIM$_{Target}$))

Condition of candidate gene for drug targets

Update Result4 = ((PharmaGKB$_{\text{Target}}$ $\cap$ DrugBank$_{\text{Target}}$) $\cup$ (FDA$_{Target}$ $\cap$ OMIM$_{\text{Target}}$))

Landmarks for optimal subways stations to connects with track(pathways both cancer and non- cancer)

Corollary-1:   IF((HubStation$_{\text{FDA}}$) $\subset$ (HubStation$_{\text{OMIM}}$)) $\cap$HubStation$_{\text{FDA}}$ =

Non-cancer disease && IF(Result3 = Non-cancer disease)

Update

Result4 = alternate Disease from drug combination

Result5 = alternate signature genes

Step2: *Input:* Drug List, Result4, Result 5

Output: alternate gene list, orphan and rare disease

Fed Query(Drug list, Result5) $\rightarrow$ (DGIdb, dGENE, TARGETdb)

Update Result5 = Gene List(Substation connected with drug interactions)

Var Result6 = Alternate Drug Name (Connection to alternate route of subway based on treatment of drugs)

If(fed query()$_{\text{Target}}$) $\subset$ Result5$_{\text{FDA}}$ = ϕ

then, Query(CancerDR) $\rightarrow$ Update(Result5) = Approved cancer Target (Established subway station) Result 6 = Alternate Tissue/Cancer Type.

If(Result5 $\subset$ Result2) && Result6 = Cancer

then, Query(Drug List, Result5) $\rightarrow$ CTD

var Result 7 = KEGG, REACTOME Pathway (Connection of two subway station on secured track) && Var Result 8 = After Another Disease

If(Result 8 $\neq$ Cancer type)

Query(drug List, Result 5) $\rightarrow$ (Orphanet, Cancer Resource, Cancer Commons)

Update Result8 = Alternate Cancer

Var Result 9 = Orphan & rare disease.

Step3:

Input: Result 5

Output: Mutation, Expression (Reason of Breakdown in Subway connection)

Fed Query(Result5) $\rightarrow$ (ClinVar, ICGC, TCGA, COSMIC, CiVIC, DoCM)

Var Result10 = Somatic Mutations, Alternate pathways, pathways specific to cancer type, clinical validations and histology

Step4:

Input: drug list, Result 5

Output: Connection and disjoint pathways

FedQuery(drug list, result 5) → (KEGG, REACTOME, WikiPathways)

Var Result11 → Cancer Process Pathways (Subway Jammer)

Var Result 12 → non-cancer non-disease pathway(subway station alternate paths)

Var Result 13 → Other Disease Pathways(Disjoint station link)

Query(drug list, result 5) → Pathway commons

Update, Result 11 → Interactome cancer

Update Result 12 → Interactome non-cancer

Update Result 13 → Interactome rare disease

Step5:

Input: Result11, Result12, Result13, Result 5, drug list

Output: Dosage Information

Query → Sider

var Result 14 → drug side effects(cost of alternate subway path)

Query(Result 11, Result 12, Result 13) → (BioModles && GO)

Var Result 15 → (Biological process & Connection with known cancer stations) → Overall effect on station administration

Query(Result5) → BindingDB

var result 16 → compound & adverse effect on pathways.

Query(drug list) → cancerrxgene

var Result 18 → Dosage Distribution, IC-50, IC-90, screening concentration, cell lines treatment

Step for compedidum of meaningful result:

Merge(Result5, Result11, Result12, Result13, Result14, Result15, Result16, Result17, Result18) → Key Targets, Key Pathways, Key Drug, Kay Alternate Indication, Key IC-50 & IC-90, key interactions.

4.3 Data Complexity

Our approach is based on the principle of integrating data from multiple resources regardless of the distribution of data. A total of 25 databases has been identified for the purpose of integration. Most of the databases are already in the form of linked data. We have used our in-house RDFizer tool to convert the non-RDF databases to RDF (linked data). Table 1 provides detailed statistics of all the databases used in this paper. The first column represents the database name and the rest of five columns represents the number of triples, number of subjects, the number of predicates, the number of objects and size of the RDF data, respectively. As evident from the table data volume is a major challenge for managing such data. The total count for a number of triples is calculated as 2.6 billion, sizing 80 GB. Only COSMIC-Mutations database has more than 1 billion triples, sizing 11.5 GB. Querying such large data from multiple sources and combining results to get useful insight is also challenging.

Table 1. RDF data statistics

Dataset	Triples	Subjects	Predicates	Objects	Size
DrugBank	3672531	316950	105	370346	594
BioModels	2399591	192582	207	1120577	405
ClinicalTrials	98835804	7337123	168	7565495	8535
GOA	28058541	5950074	36	6575678	5858
KEGG	50197150	6533307	141	6792319	4302
OMIM	8750774	1013389	101	1415364	1479
Orphanet	377947	28871	38	42891	231
PathwayCommons	5700724	1024572	69	962811	835
PharmaGKB	278049209	25325504	88	25684235	31185
Sider	17627864	1222429	39	648555	2186
WikiPathways	514397	71879	20	58611	122
FDA	2353207	102895	21	708334	63 MB
Reactome	12471494	2465218	237	4218300	957
CancerDR	NA	NA	NA	NA	NA
TARGETdb	1000	272	8	816	120 KB
DGENE	32426	11025	21	23484	3.8 MB
DGIdb	NA	NA	NA	NA	NA
ctd	151485732	13627566	27	14136295	NA
CancerResource	NA	NA	NA	NA	NA
SuperTarget	NA	NA	NA	NA	NA
CancerCommons	NA	NA	NA	NA	NA
cancerrxgene	NA	NA	NA	NA	NA
BindingDB	NA	NA	NA	NA	NA
ClinVar	NA	NA	NA	NA	NA
ICGC	NA	NA	NA	NA	40 GB
COSMIC-Mutations	1276879839	152605444	51	158166523	11.5 GB
CIViC	NA	NA	NA	NA	NA
TCGA	639344801	31019278	185	73815395	10.5 GB
DoCM	NA	NA	NA	NA	NA
Total	2.6 B	0.25 B	1562	0.3 B	120 GB

5   Results and Discussion

Identify the dosage distribution for 25 known approved cancer drugs dissecting Druggability, Omics signature, Interactome, Pathways, alternate indications and adverse events to connect disjoint subway stations problem: We have quarried in 25 anticancer drugs with our 5-layer

Table 2. Pathway Driven Drug Selection results-1

Entity	Result
Drugs(25 Anti-Cancer Drugs)	Paclitaxel Vinblastine Vinorelbine FARYDAK Tanespimycin Camptothecin Elesclomol Thapsigargin EpothiloneB PD0325901 Selumetinib Nutlin3 Docetaxel NVPAUY922 Saracatinib Irinotecan Topotecan Vinorelbine LBW242 Nilotinib Gemcitabine Vinorelbin Doxorubicin Methotrexate Topotecan NVP-AEW541
Key Genes Identified for this combination of Drugs	TOP1 IL17F PRB2 DAPK1 DCP1B EPHA4 PLA2G3 PISD LIMK2 PAPD7 NAT2 CD9 TRIO PSD3 DPH6 ACTL7B DGKB MTHFR ABCC4 DCBLD1 PLCB1 C8ORF34 DCK SLCO1B1 SYNGR1 APOBEC3B CACNA1I APOBEC3C GLIS3 PPM1L ARL14 LSM8 CFTR PVT1 ADCY8 CTTNBP2 CDH6 ZMIZ1 KIAA1377 ANGPTL5 CSMD1 RALYL SNX16 SLCO1A2 GABRB3 GABRG3 CMKLR1 NUAK1 C12ORF75 CORO1C FICD CHST11 STAB2 POLR3B BTBD11 RFX4 TGIF1 ANKS1B ARL1 MYL12B ANO4 SPIC MTUS2 PBX1 LMX1A C10ORF11 PTPRM ELFN2 C1QTNF6 RAC2 CYTH4 UGT1A1 ABCB1 CYP3A4 CYP3A5 GSTP1 NCKAP5 UGT1A9 CYP2C8 UGT1A7 ABCG2 ABCC2 ATIC SLC19A1 TYMS GSTM1
Bimodel and Most Frequent Gene Annotations	BIOSYSTEM:711360 138045 (BIOSYSTEM) BIOSYSTEM:755440 1019520 1017634 (BIOSYSTEM)BIOSYSTEM:1269379 1019520 1017634 (BIOSYSTEM) BIOSYSTEM:456 83048 (BIOSYSTEM), BIOSYSTEM:83067 672446 1269379 478 (BIOSYSTEM) BIOSYSTEM:952859 83105 373889 799177 83091 1269170 137958 198759 502 1269379 373901 1019520 960764 1269203 137928 799197 1017634 (BIOSYSTEM) BIOSYSTEM:755436 (BIOSYSTEM) BIOSYSTEM:712093 (BIOSYSTEM) BIOSYSTEM:198845 456 83061 83105 83067 1059539 471 1269379 1019520 83106 83081 1084232 198779 198827 83048 492 518 478 198795 1017634 (BIOSYSTEM) BIOSYSTEM:219801 198806 (BIOSYSTEM) BIOSYSTEM:198810 1269379 137963 198782 198774 (BIOSYSTEM) BIOSYSTEM:138045 (BIOSYSTEM), OMIM:126420 (OMIM) BIOSYSTEM:711360 (BIOSYSTEM) BIOSYSTEM:1059539 1269379 1084232 (BIOSYSTEM),GO:0090484 GO:0043225 GO:0072341 GO:0050649 GO:0008509 GO:0042802 GO:0034875 GO:0033695 GO:0042626

(Continued)

Table 2. *(Continued)*

Entity	Result
Alternate Drug	methotrexate, irinotecan
Alternate Disease	Blood group, junior system Inflammatory bowel disease 13 Homocysteinemia due to MTHFR deficiency Ivacaftor response Congenital bilateral absence of the vas deferens Crigler-Najjar syndromeEPILEPSY, CHILDHOOD ABSENCE, SUSCEPTIBILITY TO, 5; ECA5 GILBERT SYNDROME INFLAMMATORY BOWEL DISEASE 13; IBD13 DIABETES MELLITUS, NEONATAL, WITH CONGENITAL HYPOTHYROIDISM HOMOCYSTINURIA DUE TO DEFICIENCY OF N(5,10)-METHYLENETETRAHYDROFOLATE REDUCTASE ACTIVITY HOLOPROSENCEPHALY 4; HPE4 HYPERBILIRUBINEMIA, TRANSIENT FAMILIAL NEONATAL CRIGLER-NAJJAR SYNDROME, TYPE II
Pathways	C058309 C026209 C060533 D003022 C025483 D013442 C029350 C033146 C059141 C068337 D012293 D000927 C107497 C082032 C086276 D013148 C011864 C028473 D013988 C051890colocalization by mmunostaining(M0025) anti bait coimmunoprecipitation(M0089) tandem affinity purification(M0088) chromatography technology(M0085) fluorescence technology(M0052) Invitro(M0084) elisa(M0051) Biochemical Activity(M0083) drug target gene(M6001) Gel retardation assays (M0019) molecular sieving(M0018) surface plasmon resonance(M0049)

querying model. And summary of top results is compiled in Tables 2 and 3. It's evident that we were able to identify certain genes which are different from target seed genes for each drug. These genes helped us to identify island exists between the tracks and unknown landmarks. The key aspect of this relationship between gene and drug could be understood by Fig. 5 where the overall interaction have been divided into two clusters where the larger cluster identifies the pathways connectors for Subway model which infers the enriched genes having the association with pathways alteration and alternate indication. Whereas the smaller cluster defines the core group of genes for cell stability in cancer and least affected by change of IC-50 values. The salmon color is the key pathways connectors for each subway station and loss of these may end up alternate function or dysfunction of a drug.

Now using Linked Functional annotation developed using these set of databases [23] we have identified that for cancer drugs the key IC-50 value ranges from -3.050569 to 6.767348. Whereas for the alternate disease, it should be -3.978497 to -6.274387. The alternate associations with diseases such as in Retinitis pigmentosa, Chronic mucocutaneous candidiasis, Candidate gene tested

Table 3. Pathway Driven Drug Selection results-2

Entity	Result
Genomic Region	12q23.3 22q13.1 12q24.1 2q37 7q21.1 15q12 22q12.2 11q22.1 12q23.2 8q24 4q13.3-q21.1 12q23.1 22q12.3-q13.1 7q31.1-q31.3 10q22.3 22q13.1-q13.2
Mutations	Breast cancer patient xenografts (British Columbia, Nature 2014) Cancer Cell Line Encyclopedia (Novartis/Broad, Nature 2012) Skin Cutaneous Melanoma (TCGA, Provisional) Lung Squamous Cell Carcinoma (TCGA, Provisional) Desmoplastic Melanoma (Broad Institute, Nat Genet 2015)ABCB1 ABCC2 ABCC4 ABCG2 ACTL7B
Interactions	VINBLASTINE and TUBB2A VINBLASTINE and TUBB VINBLASTINE and TUBA1A VINBLASTINE and TUBE1 TANESPIMYCIN and HSP90AB1 TANESPIMYCIN and HSP90AA1 CAMPTOTHECIN and TOP1 EPOTHILONE B and TUBB EPOTHILONE B and TUBB1 EPOTHILONE B and TUBA1C EPOTHILONE B and TUBA4A SELUMETINIB and KRAS SELUMETINIB and MAP2K1 SELUMETINIB and MAP2K2 SELUMETINIB and BRAF DOCETAXEL and XRCC1 DOCETAXEL and IGF2 DOCETAXEL and CXCR4 DOCETAXEL and FOLH1 NILOTINIB and KIT NILOTINIB and ABL1 NILOTINIB and PDGFRB NILOTINIB and PDGFRA NILOTINIB and UGT1A1 GEMCITABINE and XRCC1 GEMCITABINE and WEE1 GEMCITABINE and MAGEH1 GEMCITABINE and ZEB1 DOXORUBICIN and ZEB1 DOXORUBICIN and TP53 DOXORUBICIN and LRP1B DOXORUBICIN and TOP2A
Approval and CT status	Most of the drugs are in recruiting and clinical trails stage 3 or 4 if the drugs are in the recruiting stage
Alternate Cancer types	lung_small_cell_carcinoma kidney melanoma head_and_neck Bladder lung_NSCLC_adenocarcinoma pancreas Myeloma oesophagus neuroblastoma acute_myeloid_leukaemia chronic_myeloid_leukaemia lung_NSCLC_squamous_cell_carcinoma lymphoblastic_leukemia large_intestine lymphoblastic_T_cell_leukaemia endometrium biliary_tract skin_other Burkitt_lymphoma lung_NSCLC_large_cell B_cell_lymphoma
Key Cell Processes	Apoptosis regulator Bcl-2 Cytochrome P450 19A1 Cytochrome P450 1B1 Cytochrome P450 2C8 Cytochrome P450 2C9 Cytochrome P450 3A4 Cytochrome P450 3A5 Cytochrome P450 3A7 Tubulin beta-1 chain
Cell-lines	KYSE-150 TE-10 TE-12 CTV-1 Daudi HT NALM-6 OPM-2 OPM2 PF-382
Side effects	Alopecia Diarrhoea Leukopenia Nausea Pain Agranulocytosis Neutropenia Gastrointestinal disorder Lymphopenia Amenorrhoea Alopecia Mucosal inflammation
Rare and Orphan Disease	Disease-causing germline mutation(s) in?Retinitis pigmentosa Chronic mucocutaneous candidiasis Candidate gene tested in?Amyotrophic lateral sclerosis
Dosage Details[High Confidence]	−3.050569, 6.767348, −3.978497, −0.567584, −0.861982, −4.178345, −4.85306, −6.274387

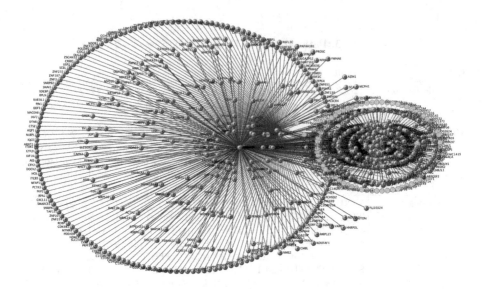

Fig. 6. Pathway Associations and seed genes for drug targets

in Amyotrophic lateral sclerosis are notable outcomes. An alternate drug such as methotrexate, irinotecan in combination helps to understand many pathways and one of the key outcomes of Subway model. Lastly Blood group, junior system, Inflammatory bowel disease 13 are key cancer types dined out using linked database approach. We have reported detailed result from this paper in Tables 2 and 3.

6 Future Work

This work will be extended to cover the knowledge graph-based modeling of presented Subway-Highway model. Re-ranking of results based on knowledge graph embedded of concepts such as pathways, mutations etc. will help to understand the relevance of each drug combinations. Adding the features such as suggestion relatedness in cancer genomic specific knowledge graphs will help to identify entry and exit point in subway model. These related entry and exit points will define alternate pathways and indications derived from input drugs. Implementing this relatedness with models such as RECAP approach will help interlinked entities to redefine their occurrences for each drug type [25]. Further results will be validated again with various ranking algorithms after embedding genomics entities into single knowledge graph against the state of the art. Present approach retains a proof of concept with limited RDF data sets and all proposed datasets will be covered in future. In the present paper we have also used non-RDF datasets to obtain results.

7   Conclusion

The present paper demonstrates a key Subway model to query and mining data associated with combinations of the drugs in cancer. Results explain that integration of databases in a multi layer structure will provide dosage balancing, alternate indications based on modifications in the pathways.[3]

Acknowledgments. This publication has emanated from research supported in part by a research grant from Science Foundation Ireland (SFI) under Grant Number SFI/12/RC/2289.

References

1. Kumar, R., Chaudhary, K., Gupta, S., Singh, H., Kumar, S., Gautam, A., Kapoor, P., Raghava, G.P.: CancerDR: cancer drug resistance database. Scientific reports 3 (2013)
2. Gottlieb, A., Altman, R.B.: Integrating systems biology sources illuminates drug action. Clin. Pharmacol. Ther. **95**(6), 663 (2014)
3. Hahn, W.C., Weinberg, R.A.: A Subway Map of Cancer Pathways. Nature Publishing Group (2002)
4. Spanheimer, P.M., Cyr, A.R., Gillum, M.P., Woodfield, G.W., Askeland, R.W., Weigel, R.J.: Distinct pathways regulated by RET and estrogen receptor in luminal breast cancer demonstrate the biological basis for combination therapy. Ann. Surg. **259**(4), 793 (2014)
5. Croset, S., Drug repositioning and indication discovery using description logics (Doctoral dissertation, University of Cambridge) (2014)
6. Joshi-Tope, G., Gillespie, M., Vastrik, I., D'Eustachio, P., Schmidt, E., de Bono, B., Jassal, B., Gopinath, G.R., Wu, G.R., Matthews, L., Lewis, S.: Reactome: a knowledgebase of biological pathways. Nucleic Acids Res. **33**(suppl 1), D428–D432 (2005)
7. Whirl-Carrillo, M., McDonagh, E.M., Hebert, J.M., Gong, L., Sangkuhl, K., Thorn, C.F., Altman, R.B., Klein, T.E.: Pharmacogenomics knowledge for personalized medicine. Clin. Pharmacol. Ther. **92**(4), 414 (2012)
8. Chen, X., Wang, Y., Wang, P., Lian, B., Li, C., Wang, J., Li, X., Jiang, W.: Systematic analysis of the associations between adverse drug reactions and pathways. In: BioMed Research International (2015)
9. Ritschel, W.A.: Handbook of Basic Pharmacokinetics (1976)
10. Patel, N., Itakura, T., Jeong, S., Liao, C.P., Roy-Burman, P., Zandi, E., Groshen, S., Pinski, J., Coetzee, G.A., Gross, M.E., Fini, M.E.: Expression and functional role of orphan receptor GPR158 in prostate cancer growth and progression. PloS one **10**(2), e0117758 (2015)
11. Hassanzadeh, O., Kementsietsidis, A., Lim, L., Miller, R.J., Wang, M.: A linked data space for clinical trials. arXiv preprint arXiv:0908.0567
12. Pauwels, E., Stoven, V., Yamanishi, Y.: Predicting drug side-effect profiles: a chemical fragment-based approach. BMC Bioinformatics **12**(1), 169 (2011)

---

[3] All predictions, dosage information and alternate indications having been reported by data driven approach only. Experiments must be conducted after wet-lab and clinical validations.

13. Zheng, H., Wang, H., Xu, H., Wu, Y., Zhao, Z., Azuaje, F.: Linking biochemical pathways and networks to adverse drug reactions. IEEE Trans. NanoBiosci. **13**(2), 131–137 (2014)

14. Ma, H., Zhao, H.: Drug target inference through pathway analysis of genomics data. Adv. Drug Delivery Rev. **65**(7), 966–972 (2013)

15. Karp, P.D., Krummenacker, M., Paley, S., Wagg, J.: Integrated pathway-genome databases and their role in drug discovery. Trends Biotechnol. **17**(7), 275–281 (1999)

16. Atias, N., Sharan, R.: An algorithmic framework for predicting side effects of drugs. J. Comput. Biol. **18**(3), 207–218 (2011)

17. Zhou, H., Gao, M., Skolnick, J.: Comprehensive prediction of drug-protein interactions and side effects for the human proteome. Scientific reports, 5 (2015)

18. Pratanwanich, N., Lió, P.: Pathway-based Bayesian inference of drug-disease interactions. Mol. BioSyst. **10**(6), 1538–1548 (2014)

19. Hardt, C., Beber, M.E., Rasche, A., Kamburov, A., Hebels, D.G., Kleinjans, J.C., Herwig, R.: pathway-level interpretation of drug-treatment data. Database, p. baw052 (2016)

20. Li, J., Lu, Z.: Pathway-based drug repositioning using causal inference. BMC Bioinformatics **14**(16), 1 (2013)

21. Guney, E., Menche, J., Vidal, M. and Barábasi, A.L., Network-based in silico drug efficacy screening. Nature Commun. **7** (2016)

22. Mehdi, M., Iqbal, A., Hogan, A., Hasnain, A., Khan, Y., Decker, S., Sahay, R.: Discovering domain-specific public SPARQL endpoints: a life-sciences use-case. In: Proceedings of the 18th International Database Engineering & Applications Symposium, pp. 39–45. ACM, July 2014

23. Jha, A., Khan, Y., Iqbal, A., Zappa, A., Mehdi, M., Sahay, R., Rebholz-Schuhmann, D.: Linked Functional Annotation For Differentially Expressed Gene (DEG) Demonstrated using Illumina Body Map 2.0

24. Jha, A., Khare, A., Singh, R.: Features' Compendium for Machine Learning in NGS Data Analysis (2015)

25. Pirró, G., Cuzzocrea, A.: RECAP: building relatedness explanations on the web. In: Proceedings of the 25th International Conference Companion on World Wide Web, pp. 235–238. International World Wide Web Conferences Steering Committee (2015)

26. Mehdi, M., Iqbal, A., Hasnain, A., Khan, Y., Decker, S., Sahay, R.: Utilizing domain-specific keywords for discovering public SPARQL endpoints: a life-sciences use-case. In: Proceedings of the 29th Annual ACM Symposium on Applied Computing, pp. 333–335. ACM, March 2014

27. Mehdi, M., Iqbal, A., Khan, Y., Decker, S., Sahay, R.: detecting inner-ear anatomical, clinical datasets in the Linked Open Data (LOD) cloud. In: Proceedings of International Workshop on Biomedical Data Mining, Modeling, Semantic Integration: A Promising Approach to Solving Unmet Medical Needs (BDM2I2015) (2015)

28. Callahan, A., Cruz-Toledo, J., Ansell, P., Dumontier, M.: Bio2RDF release 2: improved coverage, interoperability and provenance of life science linked data. In: Cimiano, P., Corcho, O., Presutti, V., Hollink, L., Rudolph, S. (eds.) ESWC 2013. LNCS, vol. 7882, pp. 200–212. Springer, Heidelberg (2013). doi:10.1007/978-3-642-38288-8_14

A Dynamic Data Warehousing Platform for Creating and Accessing Biomedical Data Lakes

Pradeeban Kathiravelu[1,2](✉) and Ashish Sharma[1](✉)

[1] Department of Biomedical Informatics, Emory University, Atlanta, GA, USA
{pkathi2,ashish.sharma}@emory.edu
[2] INESC-ID Lisboa/Instituto Superior Técnico,
Universidade de Lisboa, Lisbon, Portugal

Abstract. Medical research use cases are population centric, unlike the clinical use cases which are patient or individual centric. Hence the research use cases require accessing medical archives and data source repositories of heterogeneous nature. Traditionally, in order to query data from these data sources, users manually access and download parts or whole of the data sources. The existing solutions tend to focus on a specific data format or storage, which prevents using them for a more generic research scenario with heterogeneous data sources where the user may not have the knowledge of the schema of the data a priori.

In this paper, we propose and discuss the design, implementation, and evaluation of *Data Café*, a scalable distributed architecture that aims to address the shortcomings in the existing approaches. *Data Café* lets the resource providers create biomedical data lakes from various data sources, and lets the research data users consume the data lakes efficiently and quickly without having a priori knowledge of the data schema.

Keywords: Data lakes · Biomedical data repositories · Data integration · Data warehousing

1 Introduction

Medical research data is diverse in structure, storage format, and access protocols. It can be structured, semi-structured, unstructured, or even ill-formed. Medical data repositories may also consist of noisy and fuzzy or uncertain data. Despite the heterogeneity in their nature, the data stored in repositories still abide to various spatial and temporal relationships between them. Hence, data consumers require integrating the medical data following the inherent relationships among them for various researches. This becomes a challenging data engineering problem due to the variety of storage, messaging, and accessing protocols and approaches in the medical data.

Most contemporary solutions require a database administrator to initiate the migration of data into a warehousing environment that allows one to query and

© Springer International Publishing AG 2017
F. Wang et al. (Eds.): DMAH 2016, LNCS 10186, pp. 101–120, 2017.
DOI: 10.1007/978-3-319-57741-8_7

explore all the data at once. Given the cost of setting up such a warehouse, the typical approach is to create a unified warehouse and give researchers the ability to query and explore this data. However, the existing integration solutions tend to be too specific to a given data source or storage format, or expect the consumers to be aware of the storage formats or data schema. Hence they fall short in catering to a heavily heterogeneous landscape of precision medicine data, which typically consists of large number of small datasets.

As the medical data possess no shared interface due to the variety in storage and messaging protocols, further research and implementation is necessary to handle the data in petascale. Many standards such as HL7 FHIR [1] and *Consolidated Clinical Document Architecture* [4] are developed to exchange healthcare information electronically, offering REST APIs to data sources. In addition to the innovation in technology, we believe that it is necessary to innovate the workflow how the data is integrated and accessed, in order to cater the evergrowing complexity of this data integration problem.

Medical data are of multi-modality consisting of various medical data including, clinical data, imaging data, and features computed through image analysis, consisting of demographics, treatment, and lab data along with the images. These virtual datasets are defined as data lakes [20]. We propose *Data Café* as a data warehousing platform that helps integrate research data from disparate, multi-modality data sources and thus, create a problem or hypothesis specific virtual dataset or cohort. The created biomedical data lakes are accessed by the data consumers.

Data Café innovates a two-step workflow where the resource providers (i) create problem-specific data lakes from various independent data sources, and (ii) offer the users a schema-free access to the data lakes in the integrated data source or data warehouse to efficiently query them through the relevant APIs, without having to go through the manual labour of downloading and joining the data with a priori knowledge of the schema or the nature of the data. *Data Café* hence automates a large share of the process, mitigating the effort needed from the data consumers while reducing the repeated development and administration efforts by curating all the data to single unified data lakes.

Data Café contributes a set of enhancements to the current approaches in creating and consuming scalable biomedical data warehouse. *Data Café* (i) lets the data or resource providers offer the join-attributes, so that the consumers can specify the required attributes and search criteria without knowing the information on how the data should be joined from various data sources, (ii) lets the resource providers centrally orchestrate the data lake creation process from multiple related, but heterogeneous data sources, (iii) lets the users query, find, and consume subsets of data from the data lake without having a priori knowledge of the schema or the nature of the original data sources, (iv) leverages the schema when it is made available by the resource providers, and (v) offers very fast and efficient querying abilities by exploiting the Apache Drill [7] schema-free SQL query engine. Hence, *Data Café* attempts to enhance the usability of the heterogeneous biomedical data sources in research.

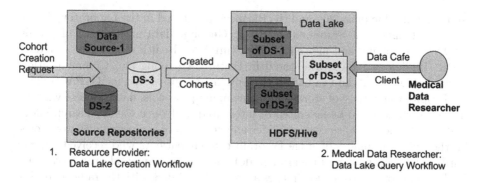

1. Resource Provider: 2. Medical Data Researcher:
    Data Lake Creation Workflow   Data Lake Query Workflow

Fig. 1. Creating and accessing biomedical data lakes

Data Café functions as a logically centralized orchestrator, creating, managing, and orchestrating the biomedical data lakes. By separating the data lake creation and consumption workflows from the original data sources that they consume, *Data Café* offers a dynamic data warehousing platform for creating and accessing biomedical data lakes, as shown by Fig. 1. It uses an unstructured scalable and logically unified distributed storage such as the Hadoop Distributed File System (HDFS) [25] to store the structured and unstructured data. The schema is stored in a distributed metadata storage such as Apache Hive [23] or in an in-memory data grid, that is integrated with the data lake. Thus, *Data Café* attempts to offer the best of the worlds, from both structured and unstructured data, in dynamically querying the biomedical data sets.

We will further analyse the proposed *Data Café* approach in the upcoming sections. Section 2 discusses the diversity in the data landscape of precision medicine, and offers the overall motivation behind the *Data Café* approach and workflow. Section 3 begins with a use case scenario and continues to discuss biomedical data lakes and how they can be constructed to give an efficient integrated access to the medical image repositories. It further discusses, in detail, the solution architecture of *Data Café*. Section 4 elaborates the prototype implementation. Preliminary evaluations on *Data Café* are discussed in Sect. 5. Related work is discussed briefly in Sect. 6. Finally, Sect. 7 concludes the paper with the current state of the research and future work.

2   Motivation

Biomedical data are stored in various health repositories and archives. The researcher needs to deduce useful information by analysing the relationships across various medical research data sources for specific research use cases.

Medical data can be clinical, imaging, or genomic data. Imaging can further be of various types such as radiology, radiation therapy (RT), or pathology. Radiology and pathology data may consist of images, features, and reports. Similarly, radiation therapy data consist of features and treatment. On the other hand, clinical data may consist of (i) Electronic Medical Record (EMR) such as treatment and demographic data [26], and (ii) Biospecimen stored in the Laboratory Information Management Systems (LIMS) [8]. Genomic data may consist of details of DNA sequencing and messenger RNA (mRNA). Health reports are stored as EMR. Radiology images are stored in Picture Archiving and Communication Systems (PACS) [9] and radiation therapy images are stored in RT-PACS [22].

Despite the variety of storage and messaging protocols, data of these varying nature often needs to be federated or integrated to deduce useful research information. As a sample use case, consider an execution of radiogenomics workflows on the diffusion images of stage IV brain tumour patients who received a standard chemotherapy drug (Temozolomide) with an experimental drug (Avastin). In this workflow, the diffusion images are stored in PACS, with the radiogenomics data in Annotations and Image Markup (AIM) [21] along with the radiation therapy and molecular genomic data. The clinical and survival data and the treatment data are stored as EMR. This workflow requires a complex data integration of heterogeneous data. *Data Café* could be used to construct a lake that integrates these diverse data sets, and allows the researchers to retrieve data for specific hypothesis such as clinical and imaging feature data of patients with poor survival and a certain mutation.

2.1 Background

Data consumers tend to consume data stored in multiple data collections inside a single database, or even data from different databases or remote data sources. Researchers either download and process the individual data sets to find patterns and solutions for interesting research problems, or use data warehousing environments to integrate the data. Joining data from multiple sources tend to be time consuming, require effort, and knowledge of the data schema a priori. Hence, often the warehousing environments are set up and provided by the resource providers. However, the current approaches offer limited capabilities to the data consumer with minimal additional features compared to manual approach. Thus the users or data consumers are still expected to be aware of the schema or specific details of the data source. Due to the limited knowledge of the data schema available to the data consumers, research efforts are necessary to improve the availability of the consolidated medical data.

Most current solutions require a database administrator to execute a batch job that replicates a data warehouse that is often constructed from multiple medical data repositories. The end users of data are finally given data on disks and/or copied to spreadsheets. This results in poor data management. Moreover, this prevents the data retrieval to be reproduced, hence resulting in various redundant duplicate data sets or locally in the data consumers computer or disks. Due to the huge size and varying nature of the biomedical archives or repositories, it is necessary to optimize how the data is integrated and accessed. Traditionally, the repositories are created by a few number of resource providers, while globally accessed by a large number of data consumers. Moreover, the data storage workflow can be less efficient as it is not in the critical path, and it is not invoked frequently. However, the data access workflow should be more efficient as it is invoked by multiple data consumers and end-users.

Having a unified warehouse with access to query and explore the data consists of its limitations when done by the data consumers themselves. Migrating large amount of data from various data sources to a unified data warehouse or a data lake is best done by the resource provider, while offering a metadata of the curated data for the data consumers to efficiently search and find the data if the stored data is in fact structured. Along with the lack of potential for scalability and extensibility to incorporate new data sources, current data integration solutions and approaches also require a priori knowledge of the data models of the different data sources.

2.2 *Data Café* Approach

Data Café tackles the limitations in the existing approaches, in terms of scalability and data accessibility. It provides researchers the ability to add new data models and sources. It does so by adopting an agile approach to creating and extending the concept of a star schema to model a problem/hypothesis specific dataset, and querying the data by extending and using the APIs provided by Apache Drill [7]. We have prototyped *Data Café* on data from PhysioNet MIMIC-III database [12], a co-clinical trial that fuses human and mouse data, as well as large synthetic datasets.

PhysioNet MIMIC-III (Medical Information Mart for Intensive Care III) [12] is an anonymized collection of health data of over 40,000 patients of critical care units of the Beth Israel Deaconess Medical Center between 2001 and 2012, which is openly available for researchers. This includes information including demographics, vital sign measurements taken at the bedside, laboratory test results, procedures, medications, caregiver notes, imaging reports, and mortality. While being available from a single location as multiple tables, PhysioNet is representative of heterogeneous databases in the biomedical research data landscape. Hence, it was chosen for the preliminary assessments of *Data Café* approach and implementation.

Cohort-specific experiments need to be reproduced across various batch jobs. Currently reproducing an experiment across the warehouse is slow, manual, and requires a large effort. By offering a unified data lake with dynamic querying capabilities *Data Café* mitigates the repeated efforts while increasing the efficiency of the researchers. Extending and leveraging a stack of in-memory data grids and distributed storage and execution frameworks, *Data Café* proposes a novel architecture in creating biomedical data lakes independent of the original data sources, and letting the users consume the data lakes efficiently with little knowledge of the original data sources, their schema, and how they are joined to create the data lake.

3   Solution Architecture

The core design of *Data Café* relies on performing joins across a multitude of datasets by identifying indices that lead from one dataset to another as join-attributes and performing a graph based intersection on them. The end result is a set of keys that identify individual data entries that not only match the requirements for the corresponding database but also matches all "WHERE" clauses across all datasets involved in the exploration. Figure 2 shows the higher level view of the *Data Café* workflows.

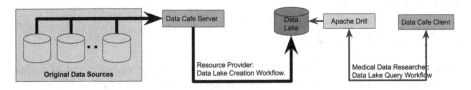

Fig. 2. Higher level view of the workflows of *Data Café*

Cohort discovery and creation is assembled per-study. Heterogeneous data is collected in a loosely structured fashion, making it agile and easy to create the biomedical data lakes. The data consumers then have problem or hypothesis specific virtual data sets. By integrating with data exploration/visualization via REST APIs, *Data Café* offers an easy access to the various relevant data sources from a single point of entry, as a data warehouse environment offered by the data providers, along with the schema and metadata information to perform the joins without the user having to know the join-attributes themselves.

3.1   Use Case

We further discuss the core *Data Café* approach and design by first going through a sample use case of real-world biomedical data repository with various data collections and multiple relationships across them. We chose PhysioNet MIMIC-III database for this simple illustration due to the fact that it is a database with a reasonable diversity and scale, that is readily available for biomedical research. Figure 3 shows a virtual data set that consists of data from 11 different datasets, each node representing a table from the MIMIC-III database. The directional edges represent a many-1 relationship through a common attribute in the respective data tables.

A query or request to create a data lake is first decomposed into a request to retrieve only the join-attributes. After retrieving join-attributes that meet a certain criteria in the tables of **ICU database definitions**, **caregiver roles**, and **ICU stays**, *Data Café* will use the data in **date formatted data** table to find a mapping between these tables. At the **admissions** node mappings between **admissions** and **date formatted data** through **admission ID** and

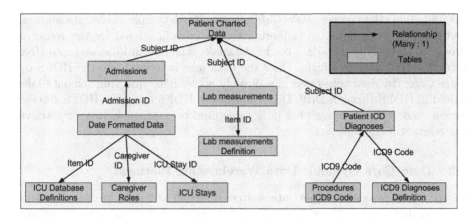

Fig. 3. Joins across a multitude of datasets from PhysioNet MIMIC-III

between **admissions** and **patient charted data** through **subject ID** will be used to perform an intersection between these nodes. **Lab measurements** links to **patient charted data** and **lab measurements definition** through **subject ID** and **item ID** respectively. Similarly, the index **ICD9 code** from **patient ICD diagnoses** is used to compute the intersection of **procedures ICD9 code** and **ICD9 diagnoses definition**. Using the **subject ID** as the join-attribute, the otherwise disjoint tables **admissions** and **patient ICD diagnoses** can be joined.

Thus the relationships across the entire table can be retrieved through the "WHERE" attributes, performed by the resource provider. Once these IDs have been retrieved, they are used to run batch retrieves that will proceed to resolve the IDs by pulling down data from their corresponding datasets. Together the data and the mapping are used as a data lake that can be used to explore this newly discovered cohort like an actual intersection has been performed across 9 different datasets. The data lake is stored in Hadoop Distributed File System (HDFS) and explored using Apache Drill.

Drill is a schema-free query engine that provides users with the flexibility to explore semi-structured, or unstructured data without having to predefine a schema. Drill further can query a range of non-relational and relational data sources out of the box, and also offer the potential to extend it through its storage plugins to integrate it with more data sources. It has a very high scalability, also with an Apache ZooKeeper [10]-based distributed execution over a cluster. As an agile platform, it offers faster insights to the queried data source. Drill is chosen as the query engine for *Data Café*, as it is an ideal query planner due to its inherent flexibility in supporting how *Data Café* stores the data lake in HDFS often in an unstructured format with schemas stored optionally in the metadata store.

The original data sources are from the resource providers, that are accessed directly through their public APIs, such as a REST API or the APIs provided by the underlying data server. *Data Café* accesses the data from these data sources, and composes the data lake adhering to the criteria offered by the resource provider. Optionally metadata or the data lake schema can be stored into Hive using the Hive storage plugin. Once the data lake is pushed into the HDFS by *Data Café*, the data consumers can directly access it by querying the data lake stored in HDFS through Drill. Drill connects to HDFS using the HDFS storage plugin, and a Hive storage plug in is configured to connect to Hive to retrieve the relevant metadata.

3.2 *Data Café* Dynamic Data Warehousing Platform

Figure 4 depicts how multiple data sources of origin and data lakes in the data layer are orchestrated through the *Data Café* server. *Data Café* server functions as the core of the control layer. Resource providers are responsible for the creation of the data lakes. The data lake creation workflow is initiated in the *Data Café* server by the resource providers. The subsets of data abiding a certain criteria or search query defined by the resource provider is copied into the data lake dynamically. During this, first the relevant IDs of the data objects are retrieved, before retrieving and copying the data with the interested attributes into the data lake as collections.

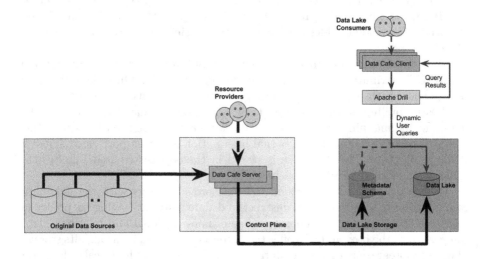

Fig. 4. Separation of concerns in the *Data Café* workflow

Each collection in the data lake may be copied from a single or multiple original data sources. In case of the structured data, the schema or metadata of the original data is leveraged in making the data access to the data lake quicker through Drill. Thus, the metadata is optionally stored in the metadata store. In the data lake consumption workflow, data lake consumers query the data lake through the *Data Café* client. The data consumers never access the original data sources; the resource providers do not have to access the *Data Café* client API that connects to the data lake through Drill.

The data lake consumers need to give only the search query and interested attributes. *Data Café* efficiently performs the join queries from the data silos or the collections from the data lakes. The graph information on how to perform the joins effectively available to *Data Café* through the metadata stored in the distributed storage provided by the in-memory data grid. Hence, *Data Café* avoids the requirement for the data lake consumers to know the schema of the data a priori. By automating this join process, the efficiency and usability of the biomedical data lakes is significantly improved compared to the state of the art research on federating medical data.

The client connects to the data lake storage and query it through a Drill cluster. When metadata or a schema is present in the optional metadata store, it will be leveraged to efficiently index and query the data stored in the metadata. The *Data Café* APIs can further be extended and implemented in the application layer for the resource providers and data consumers.

3.3   Deployment Architecture

Figure 5 indicates the deployment architecture of *Data Café* along with a sample workflow of data source integration and data query. At the step 1, the subset of data from the data repositories are queried by the resource provider and stored into HDFS as the data lake. At step 2, the data consumer uses Drill to query the data lake, which by default is hosted in HDFS/Hive in *Data Café*, but can be in any other data source such as MySQL or Amazon S3.

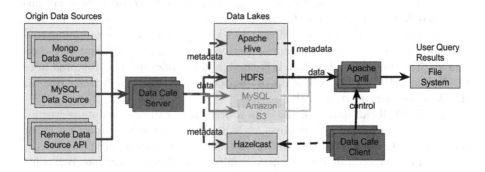

Fig. 5. *Data Café* deployment architecture

From the given set of data sources, a graphical representation of the join-attributes are created, as depicted by the example given in Sect. 3.1. This graph represents how data is connected across the various data sources. A set of parallel queries are run on the data sources that include the attributes that are present in the query graph. First relevant IDs of interest are retrieved as the result of the query.

Intersection across various join-attributes are computed. Now the data of interest can be obtained by using the ids in this intersection. A subsequent query will allow us to stream, in parallel, data from individual sources, given the relevant ids or join-attributes. This is done in two separate workflows: (i) The data lakes or integrated data source construction workflow as performed by the resource providers in the beginning; (ii) The data consumption workflows through the dynamic queries performed by the data consumers.

The metadata as well as the intermediate data objects during the data transfers should efficiently be stored in a scalable manner. In-Memory Data Grids (IMDGs) [19] offer a unified view of resources such as memory and CPU from a number of computers in a cluster. Hence, they can be leveraged to execute tasks that are heavy in memory or CPU requirements. Research and industrial use cases adopt in-memory data grids such as Hazelcast [11] and Infinispan [15] as a distributed cache to store data in-memory and execute the tasks.

Data Café is distributed with the Hazelcast, such that it can store the data objects temporarily in-memory before storing them in the integrated data source. Hazelcast distributed maps are leveraged to store the graphs on how various data sources are related, and how can they be joined for a given set of attributes from the number of the data sources. The use of Hazelcast also enables a distributed execution from multiple instances. It offers a unified view of all the computing nodes in the cluster. It further offers a large memory pool, enabling storage of larger amount of data objects in-memory, than only storing the objects in JVM.

The original data sources can be SQL data sources in relational database management systems (RDBMS) such as MySQL databases, NoSQL data sources such as Mongo data sources, and remote data source APIs. *Data Café* server has the capability to create the data lake into various structured and unstructured data sources as well. Currently, Mongo and MySQL data sources are integrated with the *Data Café* server, as a representative projects for NoSQL and SQL data sources. The integrated data lake is created, and the relevant data is stored in to the Hadoop HDFS. The metadata or the schema is stored in Apache Hive or Hazelcast in-memory data grid.

Apache Drill can query all the data sources that can be the data server for the data lake, including HDFS, Hive, MySQL, and Amazon S3. *Data Café* client connects to the Dril, while retrieving the data schema through the Hazelcast in-memory data grid. The user query results from Dril are written to the file system in a user-specified location.

Software Architecture. Figure 6 depicts the software architecture of *Data Café*. *Data Café* is maintained and built through Maven. Datacafe-core is responsible for holding the crucial data required for initiating a data cafe instance based on Hazelcast in-memory data grid. Both datacafe-server and datacafe-client depend on the datacafe-core bundle. While datacafe-server is consumed

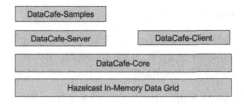

Fig. 6. Software architecture of *Data Café*

by the resource providers to build the data lakes, datacafe-client is used by the data lake consumers to efficiently access the data lakes through Drill.

Datacafe-server operates as a Hazelcast server instance, writing data into the in-memory data grid, where the datacafe-client operates as a Hazelcast client instance, merely reading the data written into the data grid. The datacafe-samples is a *Data Café* module that consists of classes that initialize datacafe-server for custom data integration tasks. It functions as a reference point to build implementations for the APIs provided by the datacafe-server, and to offer custom data integration for heterogeneous data sources.

4   Implementation

Data Café[1] has been implemented using Oracle Java 1.8.0 as the programming language. Apache Maven 3.1.1 is used for building and maintaining the source bundles. In our pilot *Data Café* deployment, we replicated data sets from the medical data repositories such as TCGA and PhysioNet and stored them locally in NoSQL data sources and relational database management systems.

We used MongoDB and MySQL to host the data from the original data repositories, as they represent the structured and non-structured database systems. MongoDB-2.4.9 and MySQL Server Ver 14.14 Distrib 5.5.41 are used as the core data sources. These data sources are consumed by *Data Café* to create the data lakes in an HDFS deployment. Data source connectors can be implemented by the resource providers to integrate other data sources with *Data Café*.

A *Data Café* client has been implemented as a layer over Apache Drill 1.6.0 to query the data lakes efficiently. Hazelcast 3.6.2 is leveraged as the in-memory data grid of the *Data Café* cluster, Apache Hadoop 2.7.2 as the data lake, and Apache Hive 1.2.0 as an optional metadata store. The bundles mysql-connector-java and drill-jdbc are used for connecting MySQL and Drill through their respective JDBC interfaces with *Data Café* respectively. Using the drill-jdbc connector, Drill is invoked through *Data Café* client.

---

[1] The source code can be found at https://github.com/sharmaashish/datacafe.

4.1 *Data Café* Modules

Data Café follows a modular architecture. We first look into the overall structure of the modules, followed by a more in-depth discussion on data sources management and *Data Café* execution.

Data Café *Client:* *Data Café* client provides a framework for the data consumers to query the data lake created by the resource providers. The data collections are stored in the data lake, each line indicating an entry, with all its chosen attributes as comma separated values (CSV) in each of the line. This storage behaviour is controlled by the datacafe.conf configuration file. For example, the file structure, delimiter, and the extension can be changed from the configuration file. *QueryWrapper* in *Data Café* client wraps the user queries from simple attributes to the search term and retrieves the interested cohorts using *DataLakeRetriever*.

Data Café *Server:* *Data Café* server provides a framework for resource providers to create the data lakes. *DataCafeUtil* consists of utility methods to query the data sources and write the data output from them into the data lake. This includes the methods to construct the query strings. Data source integrations are concurrent, and made through a separate thread per data source or a thread per data source-to-data lake data flow. *HdfsConnector* and *HiveConnector* connect HDFS and Hive to *Data Café* to host the data and metadata of the data lakes. edu.emory.bmi.datacafe.hdfs package manages the integration of *Data Café* server with HDFS and Hive. Similarly, each of the data source integration is managed by a separate package. All the connectors are multi-threaded, and also has integration with data warehouse server such that the data is written directly from the data sources to the data lakes concurrently.

If a data source integration does not exist, it can be implemented as a separate bundle, extending the *Data Café* server. The hierarchical architecture of the system makes extending the *Data Café* easy without modifying the existing code base. Moreover, JUnit test cases are implemented to test the core functionality during the Maven build or by running the relevant test cases separately.

Data Café *Samples:* *Data Café* offer samples for the application use case scenarios of *Data Café* server, and function as a framework for the developers to extend and leverage the *Data Café* server. In order to execute the *Data Café* server, an Executor should be implemented for the specific deployment, indicating the relevant data sources and configurations, as the entry point of the execution. Similarly, the consumers consume the data lake by providing their attributes for the interested cohorts, and search criteria, using the *Data Café* client. *Data Café* samples offer a starting point for the resource providers, to implement *Data Café* applications extending and executing *Data Café* server.

One notable set of samples are implemented with PhysioNet MIMIC-III database. They offer classes running the data lake creation workflow in a sequential as well as a concurrent and distributed manner, hence comparing the sequential and concurrent/distributed executions in order to evaluate the scalability offered by the *Data Café*. The samples depict integration with Mongo and MySQL data sources as well as remote repositories, where the data from the MIMIC-III database is downloaded and replicated to these data stores. Further, a simple class is implemented to start a Hazelcast instance that can join an ongoing *Data Café* cluster and contribute its memory and processing resources.

Data Café *Core:* *Data Café* core is the core enabler of distributed execution for both the *Data Café* server and the client. It is developed as a lightweight module, initializing a Hazelcast instance per an execution. The package edu.emory.bmi.datacafe.core.hazelcast consists of the classes that manage the distributed execution of *Data Café* in a Hazelcast cluster. The execution of the server instances and the client instances is distributed. The data source connectors read and write to the data sources in multiple distributed *ExecutorService* threads. The *ExecutorService* instances are sent to the remote Hazelcast instances that contain the data and operated on the data locally, than pulling the data from remote Hazelcast instances. This avoids communication overhead, enabling scalability and communication across a wide area network, such that the server can propagate metadata through an in-memory cluster to the clients. The edu.emory.bmi.datacafe.core.kernel package consists of the core utility classes and interfaces. *DataSourceRegistry* is implemented as a class to keep track of the data sources in a list. This offers auxiliary functionality to the data sources integration as a registry.

4.2 Executing *Data Café*

Data Café client and server are initialized through their datacafe.conf configuration files. Both the server and client rely on the *Data Café* core bundle. The server and client do not depend on each other, completely decoupled from each other, and distributed to different locations – *Data Café* server executed from the resource providers' clusters, and *Data Café* client executed by the data consumer. Hence, while sharing some common properties in datacafe.conf, client and server typically have different configurations that are read by the config readers.

Configurations such as (i) the host and port of the original sources and data lake, (ii) configuration information of the projects and components involved such as Hazelcast, Hadoop, and Hive, and (iii) pointers to project-specific configurations such as hazelcast.xml, are included in the datacafe.conf. *CoreConfigReader* reads the configurations, which is further extended in *Data Café* server and *Data Café* client to read the server-specific and client-specific configurations.

CoreConfigReader reads the core of the configuration file. *HzConfigReader* extends it to execute *Data Café* in a Hazelcast-based clustered scalable and distributed mode. *HzConfigReader* consists of the Hazelcast specific parameters, and a pointer to the hazelcast.conf that provides specific configurations for Hazelcast in a clustered wide area network. *HzConfigReader* is extended as *ServerConfigReader* and *ClientConfigReader* for *Data Café* server and *Data Café* client, to read the relevant server and client-specific configurations.

The executor engines are invoked at the beginning of the execution. Server and client initializations are handled by the relevant ExecutorEngine implementations. No additional custom code development is required for the data query from the data consumers except providing the exact query parameters and configurations. Configuring and initiating an execution requires no code changes to the *Data Café* server, and requires minimal developments and configurations to point to the relevant data sources and output formats.

5   Evaluation

Data Café was evaluated for its performance and features. A computer with 8 GB DDR3 memory and *Intel*® Core™ i7-4700MQ CPU @ 2.40 GHz x 8 processor was used as the platform for the evaluation. *Data Café* client was deployed along with Drill. MongoDB and MySQL data sources were configured as the data sources, with HDFS and Hive configured as the data lake and the optional metadata store respectively. When evaluating a cloud deployment, we tested it on Amazon Web Services (AWS). MongoDB was instantiated in Amazon EC2 instances, while Hive was instantiated on Amazon EMR (Elastic MapReduce). EMR HDFS was configured with 3 nodes.

5.1   Measuring Scalability of the *Data Café* Server

We replicated the MIMIC-III database and hosted it in Mongo databases, while following its snowflake schema [14]. 6 of the locally hosted MIMIC-III Mongo databases were used as the data sources in the initial evaluation. A data lake was constructed from a query across all the 6 databases, and stored the chosen subset of results from them as silos in the data lake. The simple query retrieved the complete available data of male patients with certain diseases as identified by the $ICD9\_CODE$, and created the data lake.

Figure 7 indicates the complete time taken to create the data lake from each of the data source, starting from the start-up time of *Data Café* server, till the data lake creation workflow is completed. Each data source to data lake write was handled by a single thread, hence resulting a concurrent execution of 6 workflows copying data from MongoDB to HDFS, while storing the schema in Hive metadata store. Even when there was a significant increase in the size of the data source as indicated by the number of records in the data source, the data integration workflow scaled and handled 5 million data records with up to 20 attributes well.

In order to scale the experiments and better understand the relationship between the size of the data and the data lake construction time, larger data were synthesized from the MIMIC-III databases, while still following the snowflake schema. Size of the data sources is measured by the number of records in the largest data source, where the number of attributes is kept unchanged. Figure 8 depicts the time taken to construct the data lake. It indicates that up to 500,000 records

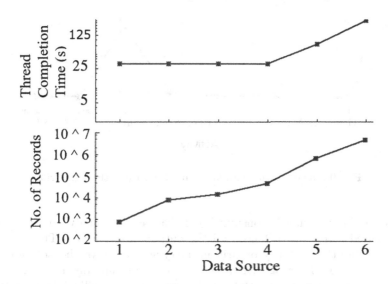

Fig. 7. No. of records in MIMIC data sources and thread execution completion time

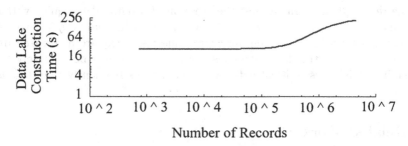

Fig. 8. Data lake construction from the synthetic data of the data sources

the time taken remain almost near constant. This initial time depends on initializing the cluster and connectors, than heavily depending on the data itself. However, beyond that, the data lake construction time increases with the data. *Data Café* server scales well to large scales of data. Furthermore, by leveraging an in-memory data grid *Data Café* can be executed faster for even larger data sets.

5.2   Evaluating Query Performance of the *Data Café* Client

The data lake was then queried using Drill by the *Data Café* client. Figure 9 indicates the time taken to complete the Drill execution and retrieve the query outcomes. The block maps were retrieved by Drill concurrently for each of the silo in the data lake. Hence 6 parallel executions with one being more time consuming than others were observed in the phase of getting the block maps from the data lake. The time consuming operation was the thread responsible for getting the block maps of data source 6, which is the largest in size, as indicated in Fig. 7.

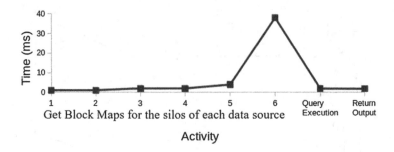

Fig. 9. Querying the data lake with *Data Café*, through Drill

However, even that thread consumed less than 40 ms. Finally, the user query was executed across all the silos and the output was returned. This was very quick, showing the efficiency of Drill in querying a data warehouse. Moreover, as Drill was invoked through *Data Café* client, the results further demonstrate that the overhead imposed by *Data Café* client to invoke Drill through its JDBC connector is negligible.

The results demonstrated that the creation of a data lake is quick without prior knowledge of the data schema. Leveraging Drill, *Data Café* provides a very fast platform for executing very large queries to explore a research data warehouse. Though the data lake construction process may be relatively time consuming, it is of less critical path, as it is done less frequently than the user queries.

6   Related Work

There have been a considerable research around precision medicine data integration and federation. Due to the number of databases and collections in the medical organizations that should be joined for any practical medical research, data warehousing solutions are designed. Some medical organizations have researched and implemented data warehousing environments and approaches to cater their needs specifically. The Enterprise Data Trust (EDT) [2] is such a data warehousing solution semantically built from various internal and external data repositories and transactional systems. It has been used as a decision support system [5] at the Mayo Clinic.

Implementing specific data warehousing solution for individual use is costly to set up and maintain. Hence, generic medical data integration and warehousing solutions are on the rise. The i2b2 (Informatics for Integrating Biology & the Bedside) project [18] is a "Hive" [16] architecture, composed of various loosely coupled cells offering multiple functionality. Creating research data sets is a primary task accomplished by one of the i2b2 cells. The data set creation process includes primary data collection and cohort identification. The curated data remain in research-specific silos, while undergoing data cleaning and integrity check workflows. Clinical Research Chart (CRC) [17] is an i2b2 cell that stores consistent metadata along with the collection of software services into a clinical research framework, by leveraging the core cells of i2b2. Due to the decoupled architecture, the remote i2b2 cells can also be replaced by locally built cells. While i2b2 does offer capabilities of data curation, federation, integration, and cleaning, it has limitations in the diversity of the data sources that it caters.

Google BigQuery [24] and Amazon Redshift [6] offer data warehousing and analytics. Improving on their approach, *Data Café* focuses on enabling easier data loading into the warehousing platform while offering fast querying capabilities. Enterprise data integration solutions such as Pentaho[2], Talend[3], Informatica[4], and CloverETL[5] enable data to be transformed and curated into warehouses from various data sources. However, they do not offer the flexibility to the data consumers in querying the warehouses through a light-weight client, or let the client join to the execution cluster with minimal resource requirements. *Data Café* can be deployed as a cluster with minimal configuration overhead and high horizontal scalability. With the schema stored separately, *Data Café* exploits the efficient distributed storage offered by HDFS and efficient SQL queries on it using Apache Drill. Fast user queries are enabled by the seamless use of the schema information from the in-memory data grid when available. The novelty of *Data Café* is in its architecture in creating warehouses quickly, while providing a client to leverage Drill with negligible overhead in querying the data stored in HDFS unstructured. This further permits the clients to leverage the schema or any other optional information stored in the in-memory cluster of *Data Café*, unlike the common medical applications or the enterprise data integration platforms.

While the common medical data integration solutions consider the complexity of the data schema, variety, and dynamics of the various data collections for the warehouse construction, they still limit themselves to the data servers that they are capable of integrating. It should further be noted that the BigData ecosystem is getting more and more diverse, and hence the current solutions that are tied to the architecture or implementation of specific data sources fail to cater the other data repositories. For example, the early warehousing and integration approaches limited themselves to relational database management systems as the unstructured data sources such as NoSQL data sources gained popularity only during the recent decades. *Data Café* attempts to be generic and data store agnostic, offering APIs to extend and implement itself even for the future data sources as discussed in the previous sections.

7   Conclusion and Future Work

Data Café is a novel platform for integrating and curating data from multiple sources into a single warehouse as data lakes. Its distinguishing characteristic is its ability to avoid the requirement of having a priori knowledge of the data

---

[2] http://www.pentaho.com/product/data-integration.

[3] https://www.talend.com/products/talend-open-studio.

[4] https://www.informatica.com/products/data-integration.html.

[5] http://www.cloveretl.com/products.

schema from the data consumers. Moreover, this also eliminates the necessity for the data consumers to access various data sources, as they are offered with an efficient and unified API built upon Apache Drill, to query the data lake. The use of indices to do the actual integration allows us to parallelize the push of the actual data into HDFS. Apache Drill provides us with a fast query execution engine that supports SQL. The SQL support is very useful, as it allows one to use it with popular business intelligence (BI) tools. In this paper we presented our first evaluation of *Data Café*. We are currently extending *Data Café* to support its deployment on to a larger multi-node distributed cluster, with multiple data stores and larger data sets. We are also putting together an IRB protocol to ingest one month's EMR data and radiology orders from the Emory Healthcare data warehouse. This data would then be used to create lakes that are representative of various clinical informatics research projects. This would also be a comprehensive and in-depth evaluation of *Data Café* capabilities and scalability.

DataScope[6] is an interactive dashboard system for doing extrapolatory analysis on large biomedical datasets. As a future work, we are planning to leverage DataScope as the front end for *Data Café*. There are ongoing efforts to evaluate the platform with diverse and heterogeneous data sources, and expanding the execution to a larger multi-node distributed cluster. Integration with imaging clients such as caMicroscope [13], as well as archives such as The Cancer Imaging Archive (TCIA) [3] and more diverse data stores and larger data repositories is being planned.

Acknowledgements. This work was supported by NCI U01 [1U01CA187013-01], Resources for development and validation of Radiomic Analyses & Adaptive Therapy, Fred Prior, Ashish Sharma (UAMS, Emory). The results shown here are partly based upon data generated by the TCGA Research Network: http://cancergenome.nih.gov/.

References

1. Bender, D., Sartipi, K.: Hl7 fhir: an agile and restful approach to healthcare information exchange. In: IEEE 26th International Symposium on Computer-Based Medical Systems (CBMS), pp. 326–331. IEEE (2013)
2. Chute, C.G., Beck, S.A., Fisk, T.B., Mohr, D.N.: The enterprise data trust at mayo clinic: a semantically integrated warehouse of biomedical data. J. Am. Med. Inform. Assoc. **17**(2), 131–135 (2010)
3. Clark, K., Vendt, B., Smith, K., Freymann, J., Kirby, J., Koppel, P., Moore, S., Phillips, S., Maffitt, D., Pringle, M., et al.: The cancer imaging archive (tcia): maintaining and operating a public information repository. J. Digit. Imaging **26**(6), 1045–1057 (2013)
4. D'Amore, D.J., Mandel, J.C., Kreda, D.A., Swain, A., Koromia, G.A., Sundareswaran, S., Alschuler, L., Dolin, R.H., Mandl, K.D., Kohane, I.S., et al.: Are meaningful use stage 2 certified ehrs ready for interoperability? findings from the smart c-cda collaborative. J. Am. Med. Inform. Assoc. **21**(6), 1060–1068 (2014)

---

[6] https://bitbucket.org/BMI/interactive-data-exporation.

5. Degoulet, P., Fieschi, M.: Medical decision support systems. In: Introduction to Clinical Informatics, pp. 153–167. Springer, New York (1997)

6. Gupta, A., Agarwal, D., Tan, D., Kulesza, J., Pathak, R., Stefani, S., Srinivasan, V.: Amazon redshift and the case for simpler data warehouses. In: Proceedings of the 2015 ACM SIGMOD International Conference on Management of Data, pp. 1917–1923. ACM (2015)

7. Hausenblas, M., Nadeau, J.: Apache drill: interactive ad-hoc analysis at scale. Big Data 1(2), 100–104 (2013)

8. Hemmer, M.: Laboratory information management systems (lims). Handbook of Chemoinformatics: From Data to Knowledge, vols. 4, pp. 844–864 (2003)

9. Honeyman, J.C., Huda, W., Ott, M., Frost, M.M., Loeffler, W., Staab, E.V.: Picture archiving and communications systems (pacs). Curr. Probl. Diagn. Radiol. 23(4), 103–158 (1994)

10. Hunt, P., Konar, M., Junqueira, F.P., Reed, B.: Zookeeper: Wait-free coordination for internet-scale systems. In: USENIX Annual Technical Conference, vol. 8, p. 9 (2010)

11. Johns, M.: Getting Started with Hazelcast. Packt Publishing Ltd., UK (2015)

12. Johnson, A.E., Pollard, T.J., Shen, L., Lehman, L.-W.H., Feng, M., Ghassemi, M., Moody, B., Szolovits, P., Celi, L.A., Mark, R.G.: Mimic-iii, a freely accessible critical care database. Scientific data 3 (2016)

13. Kathiravelu, P., Sharma, A.: Mediator: a data sharing synchronization platform for heterogeneous medical image archives. In: Workshop on Connected Health at Big Data Era (BigCHat 2015), co-located with 21st ACM SIGKDD Conference on Knowledge Discovery and Data Mining (KDD 2015). ACM (2015)

14. Levene, M., Loizou, G.: Why is the snowflake schema a good data warehouse design? Inform. Syst. 28(3), 225–240 (2003)

15. Marchioni, F.: Infinispan Data Grid Platform. Packt Publishing Ltd., UK (2012)

16. Mendis, M., Wattanasin, N., Kuttan, R., Pan, W., Philips, L., Hackett, K., Gainer, V., Chueh, H.C., Murphy, S.: Integration of hive and cell software in the i2b2 architecture. In: AMIA Annual Symposium Proceedings, vol. 1048 (2007)

17. Murphy, S.N., Mendis, M., Hackett, K., Kuttan, R., Pan, W., Phillips, L., Gainer, V., Berkowicz, D., Glaser, J.P., Kohane, I.S., et al.: Architecture of the open-source clinical research chart from informatics for integrating biology and the bedside. In: AMIA (2007)

18. Murphy, S.N., Weber, G., Mendis, M., Gainer, V., Chueh, H.C., Churchill, S., Kohane, I.: Serving the enterprise and beyond with informatics for integrating biology and the bedside (i2b2). J. Am. Med. Inform. Assoc. 17(2), 124–130 (2010)

19. Oh, J., Choi, C.-H., Park, M.-K., Kim, B.K., Hwang, K., Lee, S.-H., Hong, S.G., Nasir, A., Cho, W.-S., Kim, K.M.: Clustom-cloud: in-memory data grid-based software for clustering 16s rrna sequence data in the cloud environment. PLoS ONE 11(3), e0151064 (2016)

20. Roski, J., Bo-Linn, G.W., Andrews, T.A.: Creating value in health care through big data: opportunities and policy implications. Health Aff. 33(7), 1115–1122 (2014)

21. Rubin, D.L., Mongkolwat, P., Kleper, V., Supekar, K., Channin, D.S.: Medical imaging on the semantic web: annotation and image markup. In: AAAI Spring Symposium: Semantic Scientific Knowledge Integration, pp. 93–98 (2008)

22. Starkschall, G.: Design specifications for a radiation oncology picture archival and communication system. In: Seminars in Radiation Oncology, vol. 7, pp. 21–30. Elsevier (1997)

23. Thusoo, A., Sarma, J.S., Jain, N., Shao, Z., Chakka, P., Anthony, S., Liu, H., Wyckoff, P., Murthy, R.: Hive: a warehousing solution over a map-reduce framework. Proc. VLDB Endowment **2**(2), 1626–1629 (2009)
24. Tigani, J., Naidu, S.: Google BigQuery Analytics. Wiley, Hoboken (2014)
25. White, T.: Hadoop: The Definitive Guide. O'Reilly Media Inc., Sebastopol (2012)
26. Wilke, R., Xu, H., Denny, J., Roden, D., Krauss, R., McCarty, C., Davis, R., Skaar, T., Lamba, J., Savova, G.: The emerging role of electronic medical records in pharmacogenomics. Clin. Pharmacol. Ther. **89**(3), 379–386 (2011)

Building an i2b2-Based Integrated Data Repository for Cancer Research: A Case Study of Ovarian Cancer Registry

Na Hong[1], Zheng Li[1,2], Richard C. Kiefer[1], Melissa S. Robertson[1], Ellen L. Goode[1], Chen Wang[1], and Guoqian Jiang[1(✉)]

[1] Department of Health Sciences Research, Mayo Clinic, Rochester, MN, USA
{hong.na,li.zheng,kiefer.richard,robertson.melissa1,egoode,
wang.chen,jiang.guoqian}@mayo.edu
[2] Department of Gynecologic Oncology, The Third Affiliated Hospital of Kunming Medical University, Kunming, Yunnan, China

Abstract. In this study, we describe our preliminary efforts in building an i2b2-based integrated data repository that supports centralized data management for ovarian cancer clinical research, and discuss important lessons learnt that would inspire the evaluation and enhancement for future generic cancer-specific data repository. We collected multiple types of heterogeneous clinical data, including demographic, outcome, chemo-treatment and lab-test information for ovarian cancer. To better integrate different data types, we conducted data normalization procedures through reusing standard codes and creating mappings between local codes and standard vocabularies. We also developed the extract, transform and load (ETL) scripts to load the data into an i2b2 instance. Through further analytic practices, we evaluated major expectations of the systems according to common clinical research needs, including cohort query and identification, clinical data-based hypothesis-testing, and exploratory data-mining. We also identified and discussed outstanding issues we will address through additional enhancement of existing i2b2 system.

Keywords: Integrated data repository · Informatics for integrating biology and the bedside (i2b2) · Cancer registry · Ovarian cancer research · Extract, transform and load (ETL)

1 Introduction

Future advances in translational cancer research will be increasingly dependent on the creation of patient cohorts encompassing both highly detailed phenotypic and molecular data. Integrated Data Repositories (IDRs) [1, 2] are needed to combine molecular and phenotypic data, making data available with analytic tools, especially for cancer investigators who do not have their own computationally-focused labs. In a 2010 survey

---

Z. Li—Co-first author.

© Springer International Publishing AG 2017
F. Wang et al. (Eds.): DMAH 2016, LNCS 10186, pp. 121–135, 2017.
DOI: 10.1007/978-3-319-57741-8_8

conducted by the Clinical and Translational Science Award (CTSA) consortium [3], IDR is defined as a data warehouse integrating various sources of clinical data to support queries for a range of research-like functions. Survey results suggest that individual organizations are progressing in their approaches to the development, management, and use of IDRs as a means to support a broad array of research.

There are a number of concurrent IDR efforts, notably including the Observational Health Data Sciences and Informatics (OHDSI) [4], the National Patient-Centered Research Networks (PCORnet) [5], and the Informatics for Integrating Biology and the Bedside (i2b2) [6]. i2b2 is an open-source clinical data analytics platform that provides a component-based architecture and a flexible analytical database design. The i2b2 Star Schema was developed as a generic information model that enables transformed patient data conforming to a common data structure and representation of meaning. i2b2-based solutions have been widely used in broad clinical research communities such as the Shared Health Research Information Networks (SHRINE) [7], and the PCORnet. In the present study, we examine i2b2 for integrating a variety of clinical data sources for cancer research. We intend to use the ovarian cancer registry at Mayo Clinic as a dedicated case study to check and learn requirements in building an IDR in support of cancer-specific studies.

Ovarian cancer (OC) is the most lethal gynecological cancer. Most of OC patients undergo radical surgery to maximize reduction of initial tumor volume, and then receive at least six cycles of platinum-based chemo-treatments to eliminate remaining tumor. In a pilot study at Mayo Clinic, we are particularly interested in identifying the phenotypes related to the efficacy and side effects of the chemotherapy to potentially facilitate the in-depth patient outcome and genetic association studies. Phenotypes from the chemo-treatment period, for instance, include a serum tumor-marker named CA125 and complete blood count (CBC) test values which are routinely measured, the purpose of which is to monitor the treatment responses of patients and ensure their tolerance to chemo-induced toxicities [8, 9]. To enable this type of research, several types of clinical data are expected, including surgical records (e.g. type of surgeries, surgical outcome), chemo-treatment information (e.g. drug name, dosage), and additional laboratory test values that could be indicative of treatment efficacies and side effects.

The objective of this study is to describe our efforts in developing an IDR that supports centralized data management for ovarian cancer clinical research. We selected the i2b2 informatics platform as our testing IDR platform. We identified the datatypes (e.g., demographics, diagnoses, surgical procedures, lab tests and chemotherapy medications) and required common data elements, and enhanced default ontologies loaded in the i2b2 ontology cell. We created an extract, transform and load (ETL) tool and loaded the data that respect the data model into the data mart. We discuss the lessons learnt from the ETL process and future analytical requirements for using the i2b2 system to facilitate integrative clinical cancer research.

2 Background

2.1 i2b2 Data Repository

The i2b2 repository is a software framework allowing collaborative exchange of data including electronic health records, lab results, genetic and research data. The backend infrastructure is known as the "Hive" and is built upon a star schema which consists of facts and dimensions where facts consists of data being queried and dimensions describe those facts. The observation_fact table contains the patient data while the visit_dimension, patient_dimension, concept_dimension, and provider_dimension define the keys found within an observation_fact. Figure 1 shows the star schema of the i2b2 data model [6]. The i2b2 hive is comprised of six core cells - Project Management (PM), Data Repository (CRC), Ontology (Ont), Workplace (WORK), File Repository (FR), and Identity Management (IM). The PM cell handles user authentication along with group and role permissions. The CRC cell is the patient data star schema. The Ont cell allows for the sharing of data between institutions by managing the terminologies used within the CRC. The WORK cell manages the xml data objects used within the Hive. The FR cell stores large files such as genetic sequences and radiological images. The IM cell helps to protect patient identifiable data to satisfy HIPAA requirements. REST services are run on top of each of these cells so they may communicate with each other and external applications.

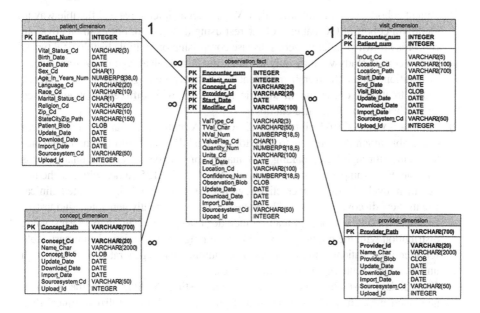

Fig. 1. The star schema of i2b2 data model

2.2   Ovarian Cancer and the Registry at Mayo Clinic

Ovarian cancer remains the most lethal gynecologic malignancy. It is the fifth leading cause of cancer death among women in the United States, with 21,290 new cases and 14,180 death in 2015 [10]. Despite of having radical surgery and initial high response rates to platinum- and taxane-based chemotherapy, most patients experience a relapse, with a median progression-free survival of only 18 months. The poor prognosis of OC patients presents an urgent need to find new treatments through investigating the disease etiology and mechanism of OC. In the present study, we intend to use an IDR-based approach to integrate ovarian cancer patients' demographic, treatment and clinical laboratory test data from different data sources, including an OC registry established at Mayo Clinic. This will facilitate discovery of novel prognosis factors and adverse effects. The OC registry mainly concerns demographic information and long-term clinical outcomes of patients for epidemiologic study; it also provides critical phenotypes and clinical outcomes for correlating with many molecular and genomic measurements [11, 12].

2.3   Functional Requirements

In this study, we intend to build an i2b2-based IDR for integrating ovarian cancer data as a use case study. The ultimate goal is to facilitate the following tasks that are commonly performed for clinical cancer research.

(i)   Patient cohort query and identification: We expect an interactive and flexible way to identify a cohort of OC patients of interest using this system. For example, ovarian cancer is known as heterogeneous disease comprising of multiple histological types, so in a certain study we may only focus on one histological type of ovarian cancer. In addition, although overall treatment is relatively homogenous across patients, actual treatment cycles might significantly differ with different platinum-based chemotherapy drugs, or different dosages.

(ii)   Exploratory analytic to facilitate new discovery: A centralized data repository system can enable much flexible way to store longitudinal data of chemo-treatment and lab-test values through time, making data analytic of unexpected clinical data patterns feasible. For example, we may analyze the trends of CA125 values within a chemo-treatment cycle. Upon an interactive review of patterns and their associated clinical events, new discovery and novel hypothesis can be potentially generated, and tested. Examples of such exploratory features include (1) a graphical summary of a test value distribution with the population-based normal range; (2) a statistical summary table for identified patient cohorts; and (3) an automatically generated KM-plot with univariate and multivariate survival analyses.

(iii)   Clinical data-driven hypothesis generation and testing: A clinical hypothesis can be generated through medical research and practices, and a centralized clinical data deposit could potentially save the efforts of linking patient information for enabling the hypothesis generation and testing. For examples, with additional lab-test values deposited in the system, we expect that it will enable an effective way to test a hypothesis such as whether the pre-surgery white-blood cell count would be associated with patient survival outcome; or whether the CA125 test values during a chemo-treatment cycle is predictive of cancer recurrence?

3   Methods

3.1   Clinical Data Collection

In total, three major data sources are used for data collection in this study.

(i)   Ovarian cancer registry database: upon study enrollment, the demographic and diagnosis data (e.g. histological types, pre-surgical CA-125) were manually retrieved and curated. The registry consists of women diagnosed or treated at Mayo Clinic who consented for research participation from 2000 to 2015. Clinical outcome data such as survival and recurrence were also routinely updated in the registry.

(ii) Lab-test database: the CA-125 and CBC (e.g., red-/white-blood cell counts) test values were retrieved from institutional lab-test database using clinical IDs of patients in the OC registry, and returned as time-stamped records with actual test values and additional test-related details.

(iii) Clinical document management (CDM) system: the chemo-treatment data of identified OC patients were extracted from chemo-treatment documents in an institutional system named CDM, which stores all patients' data about their clinical visits and medical services they received.

3.2   Build an i2b2 Instance

We chose i2b2 as an integrated data deposit as it has user-friendly query interface for clinical and biological researchers, and also it has extendable analytical modules. Our i2b2 instance was built from source code and utilizes a PostgreSQL database to store the data. Six core cells were installed into the hive. To install the i2b2 framework on a Linux server, requirements included jdk 7, jBoss 7.1.1, Ant 1.8.4, Axis2 1.6.2, cURL, and php. There are three supported databases – Oracle, SQL Server and PostgreSQL which was our selected platform. Six user accounts were created to communicate with each of the cells. After the database properties file was updated to represent our environment, scripts were run to create the tables and insert data. The services for the cells were created by configuring property files and then running a script which built the code and transferred it to the jBoss server. Once the installation was complete, the data for this project was loaded into the observation_fact table.

3.3   Load Data into an i2b2 Data Repository

We integrated heterogeneous clinical datasets into a centralized data depository in i2b2. The ETL procedures (shown as Fig. 2) comprise: (1) comparing and analyzing the i2b2 database structure and the original EMR data structure, (2) cleaning up and filtering data,

(3) creating ETL scripts to extract and transform data into a standard encoding scheme, and (4) loading data into i2b2 with de-identified patient privacy information.

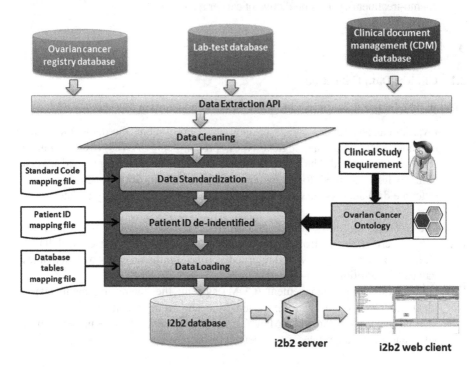

Fig. 2. The data ETL process to populate i2b2 database

3.3.1 Data Cleaning

The data cleaning process performs some basic data preprocessing, including checking and cleaning up the redundancy data, null-value data and wrong data generated by data extraction from local EMR systems. The purpose of the procedure is to ensure that data could be loaded with high quality.

3.3.2 Data Standardization

For the purpose of semantic data integration and data interaction, all the observation facts in i2b2 are standardized by a unified ontology metadata system called i2b2 Ontology. Such an i2b2 Ontology usually uses standard concept codes and the hierarchical structure of these codes together with their descriptive terms and some other information. In the i2b2 storage system, concept_dimension table contains default vocabularies terms that commonly used in clinical fields. These standard codes are mapped to specific observation facts of patients when data are imported into i2b2. Although many of the standard codes including International Classification of Diseases (ICD) [13], National Drug Code (NDC) [14], and Logical Observation Identifiers Names and Codes (LOINC) [15] are provided by i2b2 in a default installation, it still supports flexible extension when clinical study needs particular new terminology or metadata. Ontology plays a very important role in clinical and translational data analysis systems, especially when large-scale data are accumulated. i2b2 allows users to define their own metadata in its Ontology cell. We defined an Ovarian Cancer i2b2 Ontology using a bottom-up ontology building approach, i.e., through the analysis of the data we collected and clinical research questions investigators had. The i2b2 ontology is organized in the multiple dimensions to capture the ovarian cancer observation fact data extracted from different sources. Six main dimensions are defined, comprising Demographics, Diagnosis, Laboratory Test, Medication, Procedure and Vital Status. The i2b2 ontology is extensible to capture future requirements of clinical and translational study.

3.3.3   Patient De-identification

To protect patient privacy in data transformation process, we automatically allocated randomly generated patient IDs to our datasets, and maintained a mapping table for the local patient IDs and i2b2 patient IDs.

3.3.4   Data Loading

In the data loading process, we automatically transformed local data into an i2b2 compatible star schema, and loaded them into an i2b2 PostgreSQL instance database that is accessible by i2b2 client application platform. These ETL processes are written in a set of SQL scripts and Java-based scripts. When data were populated into an i2b2 database, investigators can use an i2b2 query tool to query the data, identify patient cohorts and analyze the data.

3.4   Ovarian Cancer Use Cases on i2b2 Data Repository

Once the ovarian cancer analysis data were integrated into our i2b2 platform, we invoked the i2b2 query tools and analytic plugins for patient cohort identification, patient characterization and survival analysis. In particular, we extended the i2b2 analysis ability with the R statistic computing tool[1] and performed advanced ovarian cancer data analyses such as survival analysis, through integrating i2b2 and R.

3.4.1   Patient Cohort Identification

The patient data were reviewed by our clinical research team to confirm eligibility for the i2b2 study. Patients were excluded if they (1) underwent neoadjuvant chemotherapy prior to surgery; (2) underwent prior surgery for their cancer elsewhere; (3) were treated as recurrent disease; (4) had non-epithelial or non-ovarian malignancies; (5) had no preoperative CBC parameters tested by Mayo Clinic in Rochester within 30 days prior to primary surgery. According to the patient exclusion criteria, i2b2 supports multiple groups' joint query to facilitate a reusable and one-time query running. We implemented

---

[1] https://www.r-project.org/

the above patient cohort identification in i2b2 platform by defining and employing a joint group query.

3.4.2 Patient Characteristics Statistics

The condition of patient is largely measured by the distribution of patient characteristics from different dimensions. From the clinical study requirement, we designed the measurements from 11 characteristics, which are age at diagnosis, race, origin of cancer, stage, histology, grade, preoperative CA125 level, ascites at surgery, residual disease, recurrence, and vital status. We implemented the data characterization statistics through invoking the R computing functions on the data stored in the i2b2 database.

3.4.3 Patient Survival Analysis

For advanced data analysis, we chose survival analyses as our use case and conducted the R functions on the i2b2 database. Overall survival rates were estimated via the univariate Cox proportional hazards analysis method. Based on the parameters of RDW (red blood cell distribution width), NLR (neutrophil-to-lymphocyte ratio), PLR (platelet-to-lymphocyte ratio), MLR (monocyte-to-lymphocyte ratio), combined RDW + NLR, age at diagnosis, origin of cancer, stage, histology, grade, and residual disease, the overall survival rates were calculated as the time from the date of diagnosis to the time of death. According to the results of univariate analysis, we calculated the hazard ratio (HR) and 95% confidence interval (CI). The patient survival analyses were performed using R computing functions and the ovarian cancer i2b2 instance data.

4   Results

The Ovarian Cancer i2b2 Ontology contains 6 main classes and 40 subclasses. The 6 main categories are Demographics, Diagnosis, Laboratory Test, Medication, Procedure and Vital Status as illustrated in Fig. 3. The data of a total of 1915 ovarian cancer patients enrolled in the registry and/or with retrieved test values were transformed into i2b2 tables, which respect the i2b2 Ontology we created. Using our ETL pipeline, 286,235 records of patient observation facts data with lab-test data (CA-125 and CBC) were loaded successfully.

Guided by the i2b2 Ontology, user could use built-in query tools to run queries through dragging any label from the Ontology into query tool panels. For the patient cohort identification use case we have in this study, we defined an i2b2 query to code the patient exclusion criteria outlined above. We successfully identified 654 patients in the registry (see Table 1) and with retrieved test values that meet eligibility for further patient characteristics analysis and survival analysis out of 1915 patients. Figure 3 shows the i2b2 multi-group joint query interface for our ovarian cancer patient cohort identification.

By integrating with the R statistical package, the patient characteristics statistics were computed on the identified patient cohort by 11 parameters. Figure 4 shows 2

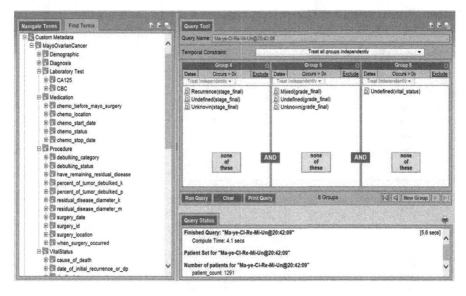

Fig. 3. An i2b2 multi-group joint query interface

examples of the descriptive statistics - Race and Vital Status distribution are displayed in the R bar graphs.

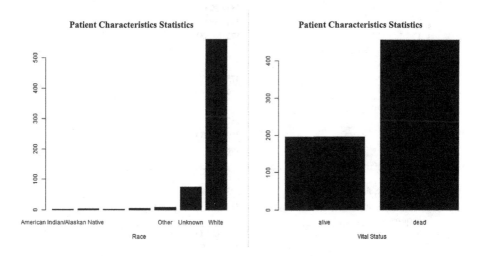

Fig. 4. Distribution of race and most recent vital status of ovarian cancer patients

For the patient survival analysis, Fig. 5 illustrated the univariate Cox proportional hazards analysis on the patient RDW using R functions and graphs. The histogram and the overall survival analysis of ovarian cancer patients (RDW cut-offs = 14.15) are

Table 1. Patient characteristics (N = 654). Note that numbers may not add to total due to missing values (75 for race, 6 for surgical debulking).

Covariates	No. of patients (%)
Age at diagnosis, years	
Mean, Range	63.0, 28–93
Race	
White	563 (97.2%)
Asian	3 (0.5%)
Other	13 (2.2%)
Origin of cancer	
Ovary	482 (73.7%)
Fallopian tube	10 (1.5%)
Peritoneum	162 (24.8%)
Stage	
I	87 (13.3%)
II	34 (5.2%)
III	416 (63.6%)
IV	117 (17.9%)
Histology	
High-grade serous	525 (80.3%)
Low-grade serous	4 (0.6%)
Endometrioid	71 (10.9%)
Clear cell	37 (5.7%)
Mucinous	17 (2.6%)
Grade	
1	28 (4.3%)
2	54 (8.3%)
3	572 (87.5%)
Preoperative CA125 level, U/ml	
<35	50 (9.5%)
≥35	475 (90.5%)
Unknown	129
Ascites at surgery	
No	187 (34.2%)
Yes	359 (65.8%)
Unknown	108
Residual disease	
None	266 (41.0%)
Macroscopic disease <1 cm	305 (47.1%)
Macroscopic disease >1 cm	77 (11.9%)
Recurrence	
No	276 (42.2%)
Yes	293 (44.8%)
Unknown	85
Vital status	
Alive	197 (30.1%)
Dead	457 (69.9%)

effectively computed and visualized with the R integration. Analyses results revealed that elevated RDW are significantly associated with patient OS (Overall Survival) with the p value < 0.001.

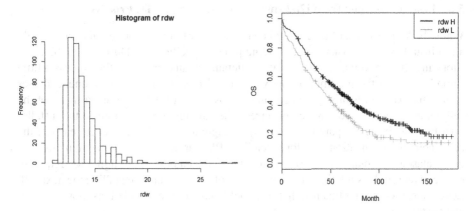

Fig. 5. Survival analysis of ovarian cancer patients and red blood cell distribution width (RDW) at diagnosis

5   Discussion

5.1   Lessons Learnt from Data Retrieval and Centralization

For medical record research not requiring informed consent, a decentralized data system usually makes it very inefficient to query and identify patient cohorts of research interest. For example, if patient eligibility criteria need to be evaluated according to multiple types of clinical data sources, it will take a lot of communication efforts and actual data retrieval steps to reach to a final cohort. If, any additional criterion needs to be reconsidered, the entire processes need to be repeated again.

While original registry database contains curated survival outcome and basic demographic data, many additional test-related and treatment-related data have not been fully incorporated due to significant amount of manual work. The examples of such data include complete blood counts (CBC) that are routinely measured before each cycle of a chemo-treatment, and detailed drug dosage data of each chemo-treatment. For the laboratory tests such as CA125 and CBCs, the internal test codes were confirmed through the lab test catalogue (http://www.mayomedicallaboratories.com/test-catalog/index.html), and the numerical values and time stamps were queried through institutional lab-test database system. The chemo-treatment data were queried and parsed from a separate clinical data management system that also contains patients' visit and billing data.

Although clinical IDs of patients have been used as a unique key to integrate different data types and link between different databases, it is critical to protect patients' privacy information by de-identifying patients when the data are used for the retrieval and integration processes. We found it encouraging that this data centralization, once accomplished, not only help reduce redundant identifiers and data mappings, but also make patient data de-identified for protecting patients' privacy.

5.2   Lessons Learned from i2b2 Installation and Data ETL Process

While the VM instance of i2b2 would be much easier to install, our team decided to do a full install as future tasks include adding project specific cells. During the i2b2 repository installation, it was found that using default parameters made the process a little smoother due to the multiple property and build files. Using the recommended user names and passwords along with running jBoss on port 9090 eliminated many of the cell communication errors which occurred when attempting to customize the installation. Even with default parameters, many configurations had to be set for each of the cells including application directory locations, PM Service URLs, connection information, datasource lookups, and jBoss path. Due to the amount of configuration required, each cell required installation then verification of success before installing the next cell. Lessons learned during this installation should make future installations more efficient and customized.

The i2b2 data model utilized a star schema that consists of table observation_fact centralized by table patient_dimension, visit_dimension, concept_dimension, and provider_dimension. In i2b2, a fact is an observation event on a patient which is not necessarily the same objects. The examples of the facts include diagnosis, procedures, health history, genetic data, lab-test data. The i2b2 dimension tables contain descriptive information about facts in different views. This data model of i2b2 plays an important role in storing and managing physical patient data on object-oriented medical ontological data model.

5.3   Lessons Learned from Ontology Building

The i2b2 Ontology we built is currently based on the use case of ovarian cancer study. It demonstrated the feasibility of the ontology in enabling the query definition for patient cohort identification and advanced analysis. However, from a long-term application perspective, the ontology building should integrate with standardized vocabularies and metadata in clinical cancer research domain, to meet the requirements of integrating cancer registry data and additional clinical data such as laboratory test data. As part of the standardization tasks in the present study, we used the mappings between local test codes at Mayo Clinic and the LOINC codes. Table 2 shows some examples of the mappings for the lab-test data. In addition, we are actively exploring to create an i2b2 ontology that integrates existing standard vocabularies such as the National Cancer Institute Thesaurus (NCIt) [16] and standard cancer metadata such as the NAACCR cancer registry data dictionary [17].

Table 2. The code mappings to support lab-test data standardization

Mapping_ID	Local_test_code	Local_test_desc	LOINC	LOINC_Test_Re sult_Name
1	28017-ROCLIS	Hematocrit, POCT, B	4544-3	Hematocrit
2	57059-ROCLIS	Hemoglobin, BF	718-7	Hemoglobin
3	9104-ROCLIS	Platelet Count	777-3	Platelet Count
4	2456-ROCLIS	Erythrocytes	789-8	Erythrocytes
5	32170-ROCLIS	Neutrophils	751-8	Neutrophils

While the i2b2 Ontology plays an important role in patient data organization and query definition, it is managed separately with the patient data in i2b2 data model. The physical patient data are associated with concept paths and identifiers. T building, management and update of the ontology are relatively independent with physical patient data management. This could not only reduce the complexity of data management, but also potentially enable a mechanism to capture the users' query needs through updating the ontology. Ideally, the ETL scripts can be invoked automatically with the updated ontology to synchronize the patient data.

5.4   Lessons Learned from i2b2 Data Analysis

Once the data are loaded into the i2b2 data repository, we successfully tested the system using the i2b2 Query Tool (a user interface) by defining a set of queries for patient cohort identification and identifying temporal patterns using i2b2 Timeline plugin. For example, we could easily perform a timeline analysis of hemoglobin lab-test of ovarian cancer patients in our data repository to capture temporal patterns of the lab-test. While the i2b2 Query tool is a critical feature that could benefit investigators to easily identify patient cohorts, investigators also desire advanced data analytic features that can be integrated with the i2b2 system. Integrating the R statistical package with the i2b2 would enable efficiency data access and empower extensible data analysis. In addition, through wrapping all the R codes as executable pipeline components, it also enables the repeatable and reusable data analysis applications. Currently, we just performed back-end integration of R and i2b2. There are a number of R plugins developed for i2b2, including R Engine Cell [18], rgate (HERON) [19], and GIRI (Generic Integration of R into I2b2) [20]. As the next step, focusing on the tasks of cancer registry and cancer study requirements, we plan to build a user-friendly i2b2 R plug-in that would allow users to configure their input and run R on top of an i2b2 web platform.

6   Conclusion

In this pilot study, we built an i2b2-based data repository system that integrates heterogeneous sources of clinical data extracted from an ovarian cancer registry and EMR systems. The query user interface and its analytic plug-ins (e.g., the timeline plugin, and R statistical plugin) are demonstrated very useful for our use cases. We have identified a number of specific analytic requirements and learned lessons for typical clinical cancer studies, which would be valuable for future i2b2 enhancements. In the future, we will also evaluate the needs to accommodate molecular and genomic measurements to achieve the full potential of an IDR like i2b2.

Acknowledgement. The study is supported in part by a NCI U01 Project – caCDE-QA (1U01CA180940-01A1), R01-CA122443, and an award from Mayo Clinic Ovarian Cancer SPORE (P50 CA136393).

References

1. Huser, V., Cimino, J.J.: Desiderata for healthcare integrated data repositories based on architectural comparison of three public repositories. In: AMIA Annual Symposium Proceedings 2013, pp. 648–656 (2013)
2. Wade, T.D., et al.: Using patient lists to add value to integrated data repositories. J. Biomed. Inform. **52**, 72–77 (2014)
3. MacKenzie, S.L., et al.: Practices and perspectives on building integrated data repositories: results from a 2010 CTSA survey. J. Am. Med. Inform. Assoc. **19**(e1), e119–e124 (2012)
4. The observational health data sciences and informatics. http://www.ohdsi.org/. Accessed 7 Mar 2016
5. PCORnet, the National Patient-Centered Clinical Research Network. http://www.pcornet.org/. Accessed 3 Mar 2016
6. Informatics for integrating biology and the bedside (i2b2). https://www.i2b2.org/. Accessed 3 Mar 2016
7. Data sharing network (SHRINE). https://www.i2b2.org/work/shrine.html. Accessed 3 Mar 2016
8. Rustin, G.J., et al.: Defining response of ovarian carcinoma to initial chemotherapy according to serum CA 125. J. Clin. Oncol. **14**(5), 1545–1551 (1996)
9. Sun, C.C., et al.: Rankings and symptom assessments of side effects from chemotherapy: insights from experienced patients with ovarian cancer. Support. Care Cancer **13**(4), 219–227 (2005)
10. Siegel, R.L., Miller, K.D., Jemal, A.: Cancer statistics, 2015. CA Cancer J. Clin. **65**(1), 5–29 (2015)
11. Konecny, G.E., et al.: Prognostic and therapeutic relevance of molecular subtypes in high-grade serous ovarian cancer. J. Natl Cancer Inst. **106**(10), dju249 (2014)
12. Wang, C., et al.: Tumor hypomethylation at 6p21.3 associates with longer time to recurrence of high-grade serous epithelial ovarian cancer. Cancer Res. **74**(11), 3084–3091 (2014)
13. International classification of diseases (ICD). http://www.who.int/classifications/icd/en/. Accessed 3 Mar 2016
14. National drug code directory. http://www.fda.gov/Drugs/InformationOnDrugs/ucm142438.htm. Accessed 7 Mar 2016
15. A universal code system for tests, measurements, and observations. https://loinc.org/. Accessed 7 Mar 2016
16. NCI Thesaurus (NCIt). https://wiki.nci.nih.gov/display/EVS/NCI+Thesaurus+(NCIt). Accessed 3 Mar 2016
17. North American Association of Centrak Cancer Registries, Data Standards & Data Dictionary, vol. II (2015). https://www.naaccr.org/StandardsandRegistryOperations/VolumeII.aspx#. Accessed 3 Mar 2016

18. Segagni, D., et al.: R engine cell: integrating R into the i2b2 software infrastructure. J. Am. Med. Inform. Assoc. **18**(3), 314–317 (2011)
19. rgate: gateway between i2b2 plugins and R (2013) https://informatics.kumc.edu/work/wiki/HeronStatsPlugins. Accessed 3 Mar 2013
20. GIRI (Generic Integration of R into I2b2) (2014). http://community.i2b2.org/wiki/display/GIRI/Home. Accessed 3 Mar 2013

Health Information Systems

AsthmaCheck: Multi-Level Modeling Based Health Information System

Tanveen Singh Bharaj[1], Shelly Sachdeva[2(✉)], and Subhash Bhalla[3]

[1] Drishti Software, Gurgaon, India
tanveensingh14@outlook.com
[2] Department of Computer Science and Engineering, Jaypee Institute of Information Technology
University, Sector-128, Noida, 201301, India
shelly.sachdeva@jiit.ac.in
[3] Department of Computer Science and Engineering, University of Aizu, Aizuwakamatsu, Japan
bhalla@u-aizu.ac.jp

Abstract. Every hospital uses their own format for creating their information system for storing patient's data. This doesn't allow hospitals to exchange patient data. Hence, there is requirement for standardized Hospital Information System (HIS). Also the HIS should be able to incorporate semantic interoperability. With technological advancement, clinicians and patients should get themselves involved in using the Electronic Health Records (EHRs). The current research provides roadmap for the introduction of domain specific clinical application following openEHR standards based on Multi-Level Modeling. Standardization will help in reducing cost and medical errors as well enhancing data quality. The current study focuses on (1) advantage of EHRs, (2) the need for standardization to improve quality of health records, thereby establishing interoperability among hospitals, (3) recognizing the use of archetypes for knowledge-based systems, (4) proposing framework for standardization, and (5) comparison of proposed approach with current HIS.

Keywords: Asthma disease · Electronic Health Records · Generic storage · Entity attribute value model · openEHR standard

1 Introduction

Chronic medical diseases take a huge toll on lives of people and these are major contributor to the rising costs in healthcare [12]. Patients nowadays are increasingly willing to take an active part in managing their conditions along with their clinicians. Therefore new applications should improve the quality of healthcare delivery. These applications often rely on recording and analyzing patient data. Thus, it is important to maintain data quality appropriately. Simultaneously, clinician's requirement for the clinical application is to get the maximum information in a limited space. They find Graphical User Interface (GUI) easier to work because their time is very limited. They need to access lot of information as fast as possible. The classical model of Hospital Information System (HIS) is based on Single-Level Modeling (SLM) where domain knowledge is

© Springer International Publishing AG 2017
F. Wang et al. (Eds.): DMAH 2016, LNCS 10186, pp. 139–154, 2017.
DOI: 10.1007/978-3-319-57741-8_9

hard-coded into the program code. The database schema created has shortcomings of maintainability and interoperability. In contrast, Multi-Level Modeling (MLM) aims to bring all systems to same standard by segregating program logic and knowledge model. Knowledge model refers to the model of health data defined and designed for clinical application. In MLM, the job of application developer is kept isolated from that of medical domain experts. Domain experts execute their task by creating small generic knowledge models called archetypes [15], rather than integrating the medical knowledge with the program logic.

1.1 Relevant Current/Open Problems in HIS

Classical HIS is built on SLM and involves lot of maintenance and redeployment. For example, endoscopy and gynecology department of a hospital will have their own database schema and corresponding individual application written in their own format. Whenever there is requirement to add new department, the entire clinical application has to go through a new iteration of development because the program logic is integrated with the medical knowledge model. The database has to be recreated and application has to be remodeled to adapt to new changes. With every new iteration, the cost of development, deployment and maintenance also increases. Also current model has issues of interoperability. Figure 1 illustrates the issue of semantic interoperability among hospitals, since every hospital build their application in their own format. The solution to stated problems is standardization. But there are certain challenges associated in adopting and implementing the standardized approach in building the clinical application. These are stated as follows:

a. *Presentation Issue*: There is a need to find a way to display health records as reliable presentation regardless of the archetypes, which are the knowledge models used to define the health data. Health records have fixed medico-legal requirement to have integrity and function in a clear and predictable manner [3].
b. *Usability Issue*: Digitalization of health records is recent. Therefore, there are more challenges in shifting to EHRs as people are not comfortable to digital health records [3]. Users are still familiar with older health record keeping software which uses old standards and approaches. Also, clinical institutions have their well-defined clinical recording practices and workflow which they are reluctant to change in order to meet new technology.

The current study aims to design and implement a prototype of a Health Information System (HIS), a standardized clinical application for a chronic disease Asthma called AsthmaCheck, following open standard specification in health informatics (openEHR). openEHR [9] is a virtual community focusing its objective on e-health interoperability and computability. Its main focus is Electronic Health Records (EHRs) and systems. The proposed application is built on Multi-Level Modeling (MLM) as opposed to Single-Level Modeling (SLM). AsthmaCheck is a standardized clinical application that supports patients with asthma.

The main design goals are usability, extensibility, and interoperability. The proposed application is capable of large data storage, incorporates scalability and also improves

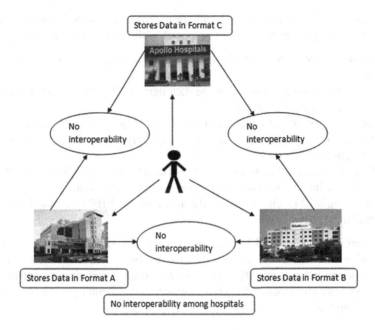

Fig. 1. Current scenario of HIS

communication between physicians and patients. It aims at initiating safer delivery, more efficient and better health care quality. Such domain specific standardized clinical application can benefit chronic disease management in several areas. The rest of the paper is organized as follows. Section 2 throws light on state of the art. Section 3 describes various aspects of proposed application. Section 4 provides implementation details. Section 5 presents results achieved. Section 6 finally concludes current work and Sect. 7 highlights future work.

2 State-of-the-Art

2.1 Problem Statement

Present scenario mostly adopts HIS that does not follow any standard. There is no semantic interoperability. Every hospital has their own way of designing their Information System (IS) based on SLM. The health information is not shareable by clinicians and among hospitals. For example, if a patient visits 'n' number of hospitals, every hospital will independently store patient's clinical information according to their IS in their own format with the respective department. For example if person 'x' is patient of Apollo hospital, the hospital can't retrieve medical history of person 'x' who has previously consulted in Fortis or Max. Apollo Hospital can only view the information in their system. EHRs on the other hand provide timely, comprehensive, and coordinated healthcare [17].

The current study is based on the openEHR approach. It's objective is to present the architecture and implementation of the proposed application (AsthmaCheck) by:

- Introducing MLM based domain specific EHR system for chronic disease (i.e., Asthma).
- Analyzing the connection between the data items in the asthma domain and their corresponding openEHR archetypes.
- Further investigating, how the data items in the asthma domain are gathered and mapped to openEHR archetypes, thus formally representing the clinical knowledge in the proposed application.
- Providing a concept for the transfer of patient data to an openEHR architecture based data storage using Postgres database called persistence of data.
- Implementing MLM architecture based domain specific EHR system and comparison with traditional EHR system.

Considering the current MLM based application, domain experts (clinicians, allied health workers, and other experts) are directly involved in defining the semantics of clinical information systems. This makes it standardized and simultaneously makes using terminology much easier [9]. Thus, it ensures interoperability and computability among all hospitals.

2.2   Single- and Multi-Level Modeling Paradigm

Most hospitals in India are constructed on classical HIS built on SLM. The database, software and GUI are developed based on an object-oriented or entity-relationship model [5]. Concepts and knowledge are encoded in the relational schema and incorporated into program code or stored procedures. In object-oriented systems, they are expressed as an object model in formalism such as Unified Modeling Language [5]. When requirements change, the existing system has to be modified and the development process enters into a new iteration of redevelopment and adds to the cost of application. For this reason the SLM results in systems which work for the present, but their utility degrades in future as they become uneconomical. In contrast, MLM is a two level approach.

- The first level of this approach is the reference model [RM]; this is a small set of classes used to support the record management functions and medico-legal requirements [19]. It comprises of terminology, definitions, identification and measurement packages [19].
- The second level is clinical model that represents the methodology of openEHR archetype [15]. Archetypes map the clinical knowledge; therefore each archetype represents one clinical concept by constraining instances of the openEHR reference model. Here the developers create generic models to be implemented on system, which on runtime are driven by archetypes (knowledge model created by domain experts).

Therefore, hard-coded software model becomes a small Reference Model (RM) capturing only semantics of health data, while domain concepts and processes are described by archetypes and then collected to generate screen forms [4]. Such information systems can evolve as needed through creation and modification of archetypes by domain experts rather than doing classical software recoding and redeployment [5]. Figure 2 illustrates multi-level modeling, which segregates tasks of application developer from that of domain experts.

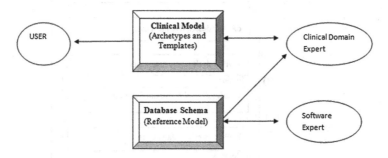

Fig. 2. Multi Level Modeling

2.3 Archetypes

Archetypes are models based on constraints of domain knowledge; they use terminologies for common language and assemble elements from RM and further constrain them [4].

The respirations archetype as shown in Fig. 3 for example represents a description of all the information a clinician may wish to report about the measurement in asthma domain. This archetype can be used in recording the details of a patient with asthma for monitoring purposes.

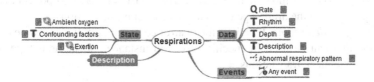

Fig. 3. Respirations archetype as mind map

Archetypes are defined by a language called Archetype Definition Language (ADL) [13]. An archetype is a data instance in openEHR that is either a *composition*, a *section* (which must be contained in a *composition* or *section*), an *entry* (which must be contained in a *composition* or *section*), or an *item structure* (which must be contained within an *entry* or *item structure*) [11]. They are stored in archetype repository.

An archetype definition basically consists of three parts as shown in Fig. 4 which describes structure of archetype file [10]. The three parts are as follows:

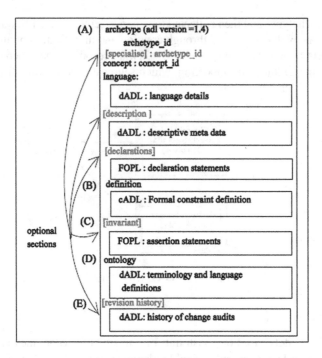

Fig. 4. Structure of archetype

(a) *Descriptive data*: Contains a unique identifier for the archetype, a machine-readable code describing the clinical concept modelled by the archetype and various meta-data such as author, version and purpose.

(b) *Constraint rules*: The constraint rules are the core of the archetype and define restrictions on the valid structure, cardinality and EHR record component instances complying with the archetype (represented by definition section).

(c) *Ontological definitions*: It defines the controlled vocabulary (i.e. machine readable codes) that can be used in specific places in instances of the archetype. It may contain language translations.

Terminologies linked in 'ontology' section aid in providing semantic interoperability of EHRs.

3 Experimentation

3.1 The Study Domain: Asthma

The current study aims to design domain specific standardized clinical application based on openEHR standard. The application caters Asthma as the domain and the application is named as AsthmaCheck. It is built on MLM.

To present a profile of introduction to the proposed application, a four step method is chosen based on waterfall model as shown in Fig. 5. In the first step, a department

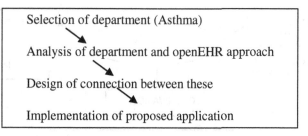

Fig. 5. Method of study

was selected. Asthma is chosen as the area of study in the development of proposed application. In the second step, a literature review was conducted for the development of archetypes.

For the analysis of the Asthma department, several meetings were performed with an expert to decide on medical knowledge and protocol to be adopted by clinicians while examining patients with asthma. In the third step, the connection was designed between the asthma department, with its different types of information gathering (systems and paper forms), and the openEHR approach. Fourth step was the implementation of the developed concepts. It was done using the programming languages such as, Java and ADL.

The proposed application is based on clinical investigator recording process [19]. The proposed application captures three types of medical information that are **'Observation'** of patients, **'Instruction'** (containing medication plan instructed to patients by clinicians) and **'Evaluation'** (containing assessment summary). These three types form **Descriptive** archetypes [17]. Since the proposed application is department specific, the 'Observation' section captures information such as clerking, history, examination and investigation along with other details. It allows clinicians to record respiration rate in / min, rhythm (regular/irregular), depth and abnormal respiration pattern along with Body Mass Index, Body Weight, Body Surface Area, and Blood Pressure as shown in Fig. 6. Archetypes used in current research are either downloaded from Clinical Knowledge Manager [8] or designed using archetype editor tool [7].

The proposed application also maintains information of various medical tests required for diagnosis and treatment of asthma. After a validity check, the patient details are permanently saved to the system in database. The clinicians can view the saved details of patients, and review individual record.

3.2   Designing Domain Specific Archetypes

There are companies such as, Ocean Informatics which offer implementation of archetype concepts and modeling tools. Archetype Editor by Ocean Informatics is an open source tool used to create and edit archetypes [7]. It has a Graphical User Interface (GUI) which is quite user friendly for semi-skilled users. It allows parsing archetypes from .adl to .xhtml and if parsed correctly, it is displayed as a GUI.

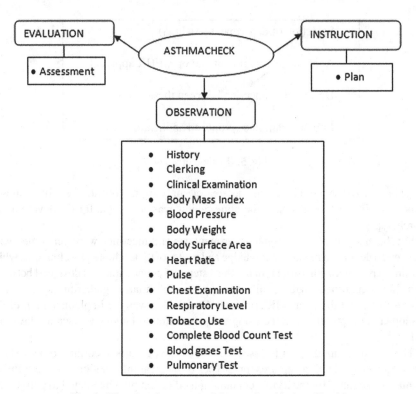

Fig. 6. List of archetypes used for creating 'AsthmaCheck'

Clinical Knowledge Manager (CKM) is a standard online library that host more than 300 archetypes related to various medical concepts. While implementing the proposed application, some archetypes required as per the medical protocol were directly downloaded from CKM and remaining ones which were not available on CKM were created and designed using archetype editor. As each archetype forms a clearly defined semantic unit that expresses one clinical concept, archetypes enable knowledge to be governed within clearly defined boundaries [10].

The current study analyses the work flow and corresponding processes in the asthma domain. Correspondingly, medical protocol for asthma examination has been designed. This protocol is shown in Fig. 7. It shows the work flow that is proposed for clinicians practicing under asthma domain. Every step represents one archetype that has been incorporated in the proposed application. The protocol illustrated is designed using resources such as Medline Plus [1], National heart, lung and blood institute [6] and review reports related to asthma protocols [16]. A domain expert has been consulted during this rigorous process of protocol development.

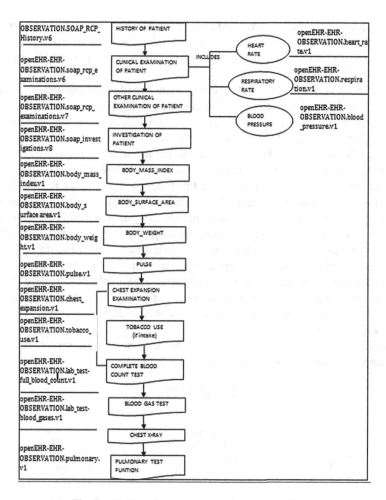

Fig. 7. Medical Protocol followed in AsthmaCheck

3.3 Architecture of AsthmaCheck (Domain Specific Clinical Application)

The working of the AsthmaCheck begins with loading archetypes into the proposed application. The proposed application runs on three important concepts inspired by project named Opereffa [14]:

(A) *GUI Generation Phase:* In the first phase of proposed application, GUI specific elements corresponding to desired archetype are built. The automatic/dynamic generation of archetype based user interfaces has been illustrated for EHRs [11, 13]. This phase requires paring .adl file of archetype to produce an archetype object model (AOM). AOM defines an archetype in form of an object oriented structure. There are set of java classes called wrappers that encapsulate function of the framework. These classes use openEHR Java Reference Implementation to provide GUI generation functionality. The Java Reference Implementation contains an

ADL parser, AOM classes and RM classes. In order to create a web form out of .adl file, there aroused a need to write code that will connect these components:

(i)   Transformation of archetypes into GUI artefacts using java ADL parser and Dojo Toolkit Framework.

(ii) Creation of RM classes, their instances and validation from data of user interface (UI) and this data must be persisted using hibernate.

After successful parsing, components of GUI are constructed considering one to one correspondence to the object contained in AOM.

(B) *Object Binding Phase:* The GUI components such as text boxes and list boxes are added to the archetypes after being parsed by DOJO toolkit [2, 14].

(C) *Data Persistence Phase:* Persistence functionality of the application is built on Hibernate, a well-known Java object relational mapping tool. Current research adopts PostgreSQL as a database management system (DBMS) for data persistence. Exiting literature [18] propose to adopt a generic schema inspired from Entity-Attribute-Value (EAV) model. Generic schema aids in capturing existing and future data without making any changes at schema level. Thus, current research is adopting basic EAV model constituting three columns namely 'Entity', 'Attribute' and 'Value'. Entity column stores primary key (patent id in our case). Attribute column reserves name of attribute whose data value needs to be stored in Value column.

Web forms constructed are connected to Apache Tomcat server that further connects to PostgreSQL database as shown in Fig. 8.

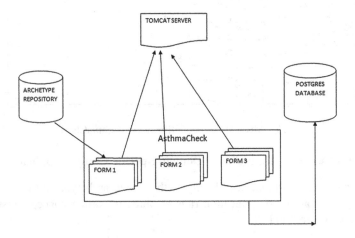

Fig. 8. Architecture of AsthmaCheck

The data values filled by user are stored in the logical database (generic in nature) which can be easily updated and viewed on demand in future. User can also easily switch between forms to make entry related to corresponding medical concept.

3.4 Detailed Working of Proposed Framework

To implement proposed application, developer firstly needs to load the required archetypes from the repository as shown in Fig. 9.

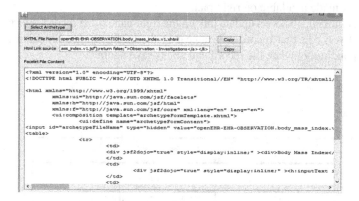

Fig. 9. Archetype producing xhtml file

The first column in the Fig. 9 contains the name of '.xhtml' file created after the ADL parser has parsed the archetype. It takes archetypes as an input and returns '.xhtml' web form as output. The second column of Fig. 9 provides with the html link source. The third column contains the parsed .xhtml file contents. Figure 10 summarizes working of the entire proposed system as explained in detailed working of framework. DOJO Toolkit adds its GUI components to the .xhtml file and these forms then get stored on Tomcat Server as illustrated in Fig. 10. To run application, certain steps need to be followed which are as follows:

Step 1:   Clinician will be asked to enter the patient name. There is change patient link also, which allows clinician to shift to another patient record by entering his name only.

Step 2:   Main application is divided into three panes. The three pane web page is the core of the UI. It is designed to provide a consistent user experience and provides patient information to clinician in single look.

The left pane contains categories such as, Observation, Instruction and Evaluation. These categories represent clinical concepts, expressed by Archetypes [9]. Each of these categories contains a hypertext list of related archetypes. The hypertext list provides mechanism to present GUI in middle pane to the user. If clinician clicks on a link, and enter data but forgets to save the form and clicks on another link then that patient details would not be persisted by the database.

Step 3:   Once the form is displayed on middle pane, clinician enter patient records for that medical domain form and clicks on save button, his data will be saved in the PostgreSQL database as illustrated in Fig. 10. This functionality is provided by data persistence phase. Entered patient details will be retrieved from database; right pane will be updated, which is again handled by data persistence phase.

Step 4: Immediately after a particular form is saved, the form in the middle pane switches to its edit mode for the last saved record of the patient with update link made available on right pane.

Step 5: If clinician clicks on update link on the right pane, for a record, a form that will let you edit that record will appear in the middle pane. Clinician can correspondingly update the record and clicks save again.

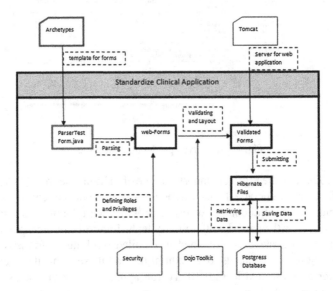

Fig. 10. Architecture proposed for implementation of AsthmaCheck

4   Implementation

4.1   Presentation

The list of archetypes parsed into '.xhtml' file appears on the left pane and represent various concepts of medical domain. The archetypes on the left pane appear as per the protocol created. The middle pane provides option for data editing and creation. Forms chosen from left pane are displayed in middle pane, where clinicians enter data that is saved in the logical database. Right pane gives overview of patient's information. After saving the data through UI, it can be viewed in right pane. The right pane also provides update option to clinicians to edit details in the form.

4.2   Data Storage

Data modeling is a creative and nondeterministic activity, which requires balancing various requirements to achieve completeness, non-redundancy, enforcement of business rules, data reusability, stability and flexibility, elegance, integration, and communication effectiveness. Figure 11 shows generic database model of proposed application.

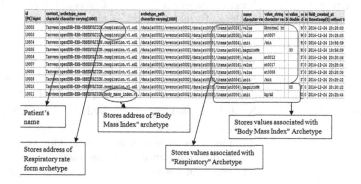

Fig. 11. The data persistence

Such database model eliminates the need of maintaining separate database for individual departments. One generic table serves data storage requirements of all departments, such as, cardiology and pathology. The table stores reference to specific department (defined through archetypes) and data values entered by clinicians in the GUI form of the corresponding department's archetype. This resolves the problem of recreating the database on requirement to add new department as in SLM.

5   Results

The system designed is usable, extensible and interoperable. The user interface developed is based on archetypes (created by domain experts). Thus, it is easy to use by clinicians, doctors and nurses. The architecture can be easily extended to other domains such as, gynecology or pediatrics department. The archetypes specific to that department/domain needs to be developed by specialists (e.g., gynecologist/pediatrician). It is based on MLM approach, which was initially proposed by openEHR, and later on adopted by ISO 13606 and HL7 (other standards aiming to achieve semantic interoperability). Interoperability among standards help in providing ability to communicate between various applications built on various standards. Table 1 compares proposed application based on openEHR standard with that of classical HIS. Archetypes make possible to leverage the vast amount of standardized terms and semantics by linking to biomedical terminologies such as SNOMED-CT [20]. The GUI based on archetypes can thus be linked to terminologies in the proposed model.

Table 1. Illustrating comparison between classical HIS and proposed model

Dimension	Classical hospital information system	Proposed solution
Addition of terminology	Large cost is incurred on addition of terminology as entire software has to be rebuilt	Edit in archetype to add terminology
Cost of building application	With every new iteration cost is added to software	Investment is made once in initial development and deployment of software
Architecture	Single Level Modeling	Multi-Level Modeling
Consistent representation	Every hospital follows their own format; therefore there is no consistent representation of information among hospitals	Consistent representation of medical information through use of archetypes which will be adopted by all hospitals
Interoperability	No	Yes
Data models implemented	Knowledge models are created by individual hospitals according to their format	Standardized knowledge models known as archetypes are created
Nature of database	Every hospital has its own database which is static specific to domain	Database is of generic nature and thus, capable of storing information of all hospitals department wise, irrespective of the fact that from which hospital the information came from

6   Conclusions

AsthmaCheck is functionally standardized clinical application based on openEHR. The GUI design has been replicated by using openEHR formalism. In the development of application we has explored the generation of user interfaces from archetypes (which are powerful representations of clinical knowledge) that parses '.adl' archetypes and correspondingly display '.xhtml' web forms. In contrast to the single level modeling paradigm, the multi-level modeling is more flexible and cuts down cost to large extent as it provides a way to handle knowledge expansion and maintains separation between domain knowledge and programming logic. The domain knowledge is created with the help of medical experts in the form of archetypes. Achieving semantic interoperability requires binding to standard medical terminologies (such as SNOMED-CT and LONIC) defined in archetype definition and coordinating archetype development through systematic "Domain Knowledge Governance" tool (CKM). This technology avoids redesigning of entire systems on addition of new terminology to archetypes.

7   Future Work

AsthmaCheck is a standardized clinical application based on openEHR which is domain specific, but we can extend this work to create a clinical application serving all or many domains in medical field. This will help in replacing entire HIS of individual hospitals with one standardized clinical application. This will ensure interoperability among all hospitals worldwide and hospitals will not require maintaining their separate database or department wise information.

Acknowledgements.   This study has been inspired from Opereffa. It stands for openEHRREFerence Framework and Application. It is a project for creating an open source clinical application which is driven by the Clinical Review Board of openEHR [14].

We would like to express our gratitude to the Dr. S. Prasad, Retired Consultant & Head of Department, Department of Medicine, Safdarjung Hospital, New Delhi. He helped us in building the medical protocol required to be followed in examining patients with asthma. The guidance helped authors in creating archetypes according to the protocol.

References

1. Asthma: MedlinePlus. http://www.nlm.nih.gov/medlineplus/asthma.html
2. DOJO toolkit. http://dojotoolkit.org/
3. Thurston L.M.: Flexible and extensible display of archetyped data: the openEHR presentation challenge. In: Proceedings of the HIC 2006 and HINZ 2006, pp. 28–36. Health Informatics Society of Australia, Brunswick East, Vic. (2006). ISBN: 0975101374
4. Atalag, K., Yang, H.Y.: From openEHR domain models to advanced user interfaces: a case study in endoscopy. In: Health Informatics New Zealand Conference, Wellington, November (2010)
5. Atalağ, K., Bilgen, S.: Multi-level modeling and the role of archetypes in the design of health information systems: a modeling example in endoscopy. In: HIBIT 2007 Proceedings of the International Symposium on Health Informatics and Bioinformatics, May (2007)
6. National Heart, Lung and Blood Institute. https://www.nhlbi.nih.gov/health/health-topics/topics/asthma/diagnosis
7. Ocean Editor Archetype. https://oceaninformatics.com/solutions/knowledge_management
8. OpenEHR Clinical Knowledge Manager. www.openehr.org/ckm
9. OpenEHR Foundation. http://www.openehr.org
10. Sachdeva, S., Bhalla, S.: Semantic interoperability in standardized electronic health record databases. ACM J. Data Inf. Qual. (JDIQ) **3**(1), 1–37 (2012)
11. Sachdeva, S., Yaginuma, D., Chu, W., Bhalla, S.: Dynamic generation of archetype-based user interfaces for queries on electronic health record databases. In: Kikuchi, S., Madaan, A., Sachdeva, S., Bhalla, S. (eds.) DNIS 2011. LNCS, vol. 7108, pp. 109–125. Springer, Heidelberg (2011). doi:10.1007/978-3-642-25731-5_10
12. Sunyaev, A., Chornyi, D.: Supporting chronic disease care quality: design and implementation of a health service and its integration with electronic health records. ACM J. Data Inf. Qual. **3**(2) (2012). Article 3

13. Schulera, T., Gardeb, S., Heardb, S., Bealec, T.: Towards automatically generating graphical user interfaces from openEHR archetypes. In: Medical Informatics Europe (MIE 2006) Ubiquity: Technologies for Better Health in Aging Societies, Maastricht, pp. 221–227. IOS Press, Amsterdam, Auguest (2006)
14. Opereffa Project. http://opereffa.chime.ucl.ac.uk/introduction.jsf
15. Beale, T., Heard, S.: Archetype definitions and principles in the openEHR release 1.0.2. openEHR Foundation (2005)
16. Dexheimer, J.W., Borycki, E.M., Chiu, K.W., Johnson, K.B., Aronsky, D.: A systematic review of the implementation and impact of asthma protocols. BMC Med. Inform. Decis. Mak. **14**, 82 (2014). doi:10.1186/1472-6947-14-82
17. Gök, M.: Introducing an openEHR-Based Electronic Health Record System in a Hospital, Master Thesis submitted to Department of Medical Informatics, University of Göttingen, Germany, May 2008
18. Patrick, J., Ly, R., Truran, D.: Evaluation of a persistent store for openEHR. In: Proceedings of the HIC and HINZ, pp. 83–89. Health Informatics Society of Australia (2006)
19. Beale, T., Heard, S.: The openEHR architecture: architecture overview. In: The openEHR release 1.0.2. openEHR Foundation (2008)
20. International Health Terminology Standards Development Organization. Systematized Nomenclature of Medicine-Clinical Terms (SNOMED CT) (2016). http://www.ihtsdo.org/snomed-ct/. (Accessed Mar 2016)

Author Index

Printed in the United States
By Bookmasters